KEN FOLEY
ERIC
PRETTY
Brian Chisholm
Bill Thompson
MARTIN KENNEDY
CLIFF ROGERS
MYRON
Lewis
Russ Grant
Howard Russell
Bill
JIM PAHAPILL
ALEX OSMOND
GREG WALSH
CINDY
RYAN
undertayker
NICK BONNETT
PAT WALSH
TED
Gerald Lush
Michael Lush
Jim
Tom
CHICO
GARY
DALTON
764
BABAR-SULTAN
MEL KEATS
GARRY RUSSELL
REID
Kyle
Sheldon way
LEN
HARRY Nolan
John Leonard
DAVID crane
Al Combs
DAN FLEMING
Keith
Norman
Henry
Voutour
HOGAN
cecil

WORKERS
LOCAL 764

Mobil Oil Canada

Photo courtesy of HMDC

The Achievement of Hibernia: The Perfect Alignment of Intent and Intended

In every company's history, there are milestone moments when expertise and unwavering effort align to realize the boldest of visions. For the past quarter century, Mobil Oil Canada has led the quest to explore for and develop hydrocarbons off Canada's east coast.

As the largest owner in Hibernia and the Sable Offshore Energy Project, and with a comprehensive portfolio of other exploration and development opportunities, we strive to provide major new sources of energy to meet the needs of North Americans well into the twenty-first century. Mobil Oil Canada salutes the dedicated efforts of our partners and employees in the achievement of Hibernia.

Hibernia

DORIS

The *Newfoundland and Labrador Oil Development Allied Trades Council (ODC)* salutes the unionized men and women of the province's construction industry on their success with Hibernia, and for paving the way to even greater opportunities for all of Newfoundland and Labrador in our emerging oil and gas industry.

Photos courtesy of HMDC

Working Together...

AOC BROWN & ROOT CANADA LTD.

Hibernia's Operations and
Engineering Alliance Partner

WITH

Accommodation Management Services

OIS-Fisher Inc.

Inspection Services

Fabric Maintenance Services

...TO DEVELOP CANADA'S RESOURCES

Photo courtesy of HMDC

Photo courtesy of HMDC

The Joint Venture between Aker Norwegian Contractors AS, a world leader in concrete platform construction, responsible for 20 Concrete Gravity Base Structures (GBS), and Peter Kiewit Sons Co. Ltd., one of North America's largest construction companies, assumed the Construction Management of the Hibernia GBS and the Marine Operations in May 1994 at the request of the project owners. We are proud that the GBS and the Marine Operations was delivered on schedule and within budget.

PASSB

Picture of Marine Base,
St. John's Harbour.

Publisher's Note

This book is a risk. It is a risk to try and capture the magnitude of Hibernia in words and photographs. It is a risk to try and convey the sense of pride in achievement experienced by the Newfoundland workers and many other workers from other countries who helped to create this wonder of the modern world. It is a risk to try to convey a sense of Hibernia's enormity—its weight (1.2 million tonnes), its height (224 metres) and the incredible fact that all of this could float! Hibernia is like a huge technological ship which amazingly can be towed through water at a speed of over three knots and safely secured on the ocean floor. It is a risk to try and capture its awesome beauty as it left its birthplace, ironically named Mosquito Cove in Bull Arm, and was towed by a flotilla of the world's strongest tugs out through iceberg-studded Trinity Bay to the fog-shrouded Grand Banks of Newfoundland. It is clearly a risk for a publisher to attempt to capture in print and illustrations the wonder of it all.

If this book succeeds it will be largely due to the focus our team adopted when we began this venture. Our focus very simply was on the worker. The men and women who built Hibernia. Their toil and sweat; their hopes and dreams and, in a broader sense, their aspirations for their children and their grandchildren. We were greatly assisted in our venture by the brilliant photography of Greg Locke and Ned Pratt whose pool of images we were kindly given access to by HMDC. The text is the work of Lara Maynard, who, following a somewhat vague and rambling description of the vision of this book I had given her, dutifully researched, wrote and edited the entire narrative to my undying gratitude. Likewise the design. My instructions to Nadine Osmond were: Give me a book that shows the people behind the project. One need only glance at what Nadine has created to sense the pride and see the look of satisfaction in these workers' faces. This was a team effort all the way and Carla Kean, Darren Hewitt, Jean Ann Rose, Michelle Cable and all the staff at Breakwater must be commended for their devotion to this project. I must also acknowledge, with thanks, the efforts of Melony O'Neill, Krista Goulding and Laura Woodford in the early stages of this project.

Like Hibernia itself, which now sits on the Grand Banks full of expectation for "first oil" we, the publishers of this book *Hibernia: Promise of Rock and Sea* await the flow of responses from our Canadian readers and others from around the world.

Best wishes,

Clyde Rose

Clyde Rose
Publisher
Breakwater Books Ltd.
St. John's
June 6, 1997

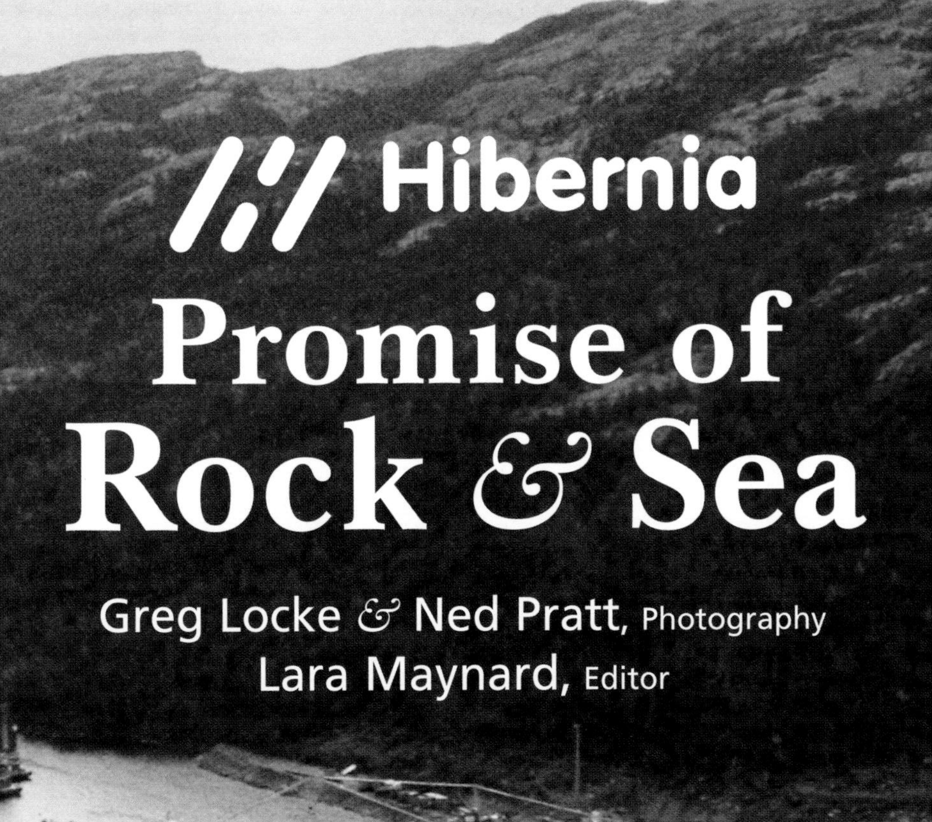

Hibernia

Promise of Rock & Sea

Greg Locke & Ned Pratt, Photography
Lara Maynard, Editor

BREAKWATER
100 Water Street
P.O. Box 2188
St. John's, NF
A1C 6E6

Canadian Cataloguing in Publication Data

Hibernia.

ISBN 1-55081-130-4

1. Hibernia Oil Field.* 2. Offshore oil industry — Newfoundland.
I. Maynard, Lara.

TN871.3.H52 1997 622'.33819'09718 C97-950099-0

We would like to thank the following for their contributions to this book: Sheila Anstey, Brian Crawley, Michelle Dawe, Patricia Jackson, Gloria Warren-Slade, the staff at the Maritime History Archive at Memorial University of Newfoundland, *Atlantic Canada Oil Works* and photographer Marc Pike.

Unless otherwise noted all black and white photography is by Ned Pratt and all colour photography is by Greg Locke.

Design and production: Nadine Osmond

This book is dedicated
to all Hibernia Project people
and especially in memory of
Bernard William Dunphy, Randolph Anthony Ducey,
Alfred March Jr. and Peter Francis Peddle

Message from the Prime Minister

It has been a long time coming, but the Hibernia project is now on the verge of production.

This is an achievement, a triumph of will and spirit, that deserves celebration. The Hibernia project is a watershed event in the annals of the oil and gas industry—an industry that has always played an integral role in Canada's economy, and in the lives of Canadians.

Hibernia has marked itself indelibly on the lives of the people of Newfoundland and Labrador and will hold a special place in our country's history. All those affiliated with the project have reason to be very, very proud of their efforts.

I am pleased that the Government of Canada has played a role in the Hibernia project. By so doing, it has actively supported development and production in the growing East coast oil and gas industry, and helped forge a new era in the region's economy.

Please accept my kindest regards.

Jean Chrétien

Jean Chrétien
Prime Minister, Canada

Message from the Premier

After years of dreaming, hoping, planning and working, Hibernia has become a reality. For Newfoundlanders and Labradorians, this massive oil and gas project signifies the beginning of a new direction for the economy, and the cornerstone of a new industry.

Hibernia: Promise of Rock and Sea, by Breakwater Books, captures the essence of construction from inception to completion. Hibernia has touched every person in Newfoundland and Labrador in some way; most elementally, the workers who poured their blood, sweat and tears into the project. Without the dedication of the team of workers, many of whom were Newfoundlanders and Labradorians, this project, touted the ninth wonder of the world, would not have reached fruition.

For the rest of us, we watched with wonder as this engineering marvel grew from a blueprint to reality. The province watched with bated breath as the GBS and Topsides were mated, and we smiled with pride as the platform reached its destination on the Grand Banks.

The official book on Hibernia holds the perpetual record of our achievements, and gives an overview of exactly how far we have come in our oil and gas development. I offer congratulations to the many people who have made this project a success.

Brian Tobin
Premier, Newfoundland and Labrador

Foreword

Hibernia Management and Development Company Ltd. is pleased to have contributed to *Hibernia: Promise of Rock and Sea.* This book is about people. More specifically, it's about the thousands of men and women who brought to life the magnificent achievement that is Hibernia.

Their pioneering effort is captured in stunning photography and each page vividly illustrates the individual contribution and sacrifice that had to be made. The text beautifully incorporates the construction workers' own words to tell this story, a story which records the faces and folklore of Hibernia for all time. Congratulations to BREAKWATER BOOKS on a job well done.

The Hibernia platform is now completed and safely in place on the Grand Banks, poised to lead the development of a new offshore oil and gas industry for Newfoundland and Labrador. This is a marvelous accomplishment. The Hibernia platform is truly a first-of-its-kind. The GBS's unique ice wall is an engineering marvel, and just one of the many safety features incorporated into the design. The quality of work which went into the Topsides, including assembly, is second to none.

What makes Hibernia even more special is that most of the work has occurred here in Newfoundland. Indeed, during peak construction more than 90 percent of the 5800 people working at Bull Arm were from the province. Hibernia's commitment to Newfoundland and Labrador will continue throughout the operations phase as well. More than 90 percent of our steady-state work force are from the province.

I extend my sincere congratulations to all who have played a role in helping Hibernia to become a success. Your contribution will not be forgotten.

This chapter of the Hibernia story is complete and we now look forward to the operations phase with renewed confidence and excitement.

Harvey Smith

Harvey Smith
President
Hibernia Management and
Development Company Ltd.

Introduction

There is more to the massive Gravity Base Structure (GBS) than meets the eye. Deep inside, several thousand tonnes of mechanical outfitting (MOF) in the form of steel structures, pipes, pumps, cables, and instruments are its heart and veins. The precious raw material, crude oil, passes through the GBS four times before it is shipped away from the platform. The reservoir pressure pushes the well stream through the Topsides for various stages of separation and stabilization before the product is piped back to the GBS for accumulation in the storage compartments. Huge crude oil pumps send the oil back to the Topsides at intervals for all the important metering prior to the fourth passage through the GBS and exit via the export pipelines to awaiting shuttle tankers. There are numerous other interesting or trivial technical details involved with the design by the MOF division, one of three engineering divisions organized and administered by Doris Engineering.

The interaction between Newfoundlanders, CFAs (Come From Aways) and CFFAs (Come From Far Aways) for almost five years in the St. John's offices should earn those several hundred technicians and engineers a mention amongst the family of "Hibernia Project People." Their job, to generate specifications, documents, and drawings for each minute detail based on visions, concepts, scientific experiments, and proven technology did establish the basis from which the industry and construction workers could proceed with actually building the structure.

After the GBS and Topsides facilities were joined, the Hibernia platform floating at Back Cove prior to the tow to the oilfield on the Grand Banks posed a truly spectacular sight, especially by night with turbine humming and lights glowing. For most people, this would be the last view of North America's first of its kind offshore production platform. The Hibernia operation workers over the next twenty to twenty-five years will be the ones to conclude that story.

Having paved the way for a new industry in Newfoundland, generated long term work-places, and extended the automotive business yet another bit, there has been one question pertaining to Hibernia many times asked: What is to come of this mighty structure when the oilfield dries up? Those who imagine future archaeologists wondering about the uncharacteristic formation on the Atlantic seabed will be disappointed. Systems, ways, and means are in place to remove and dispose of the structure after termination of its industrial life. However, a pile of pebbles makes a good home for lobsters.

The choice of the project name "Hibernia," the old name for Ireland, combined with the saying "A long way takes a strong will" (the Irish or Hibernian version of "If there is a will, there is a way"), seems to encapsulate the immense amount of willpower expended over the years by those who fought to bring the project to where it is today. And I conclude with a reflection on departing Newfoundland with its deeply rooted Irish culture: May all good Newfoundlanders forever prosper from the Hibernian heritage.

Palle E. Schwartz, "CFFA" (Denmark)

"Mankind worship success but think too little of the means by which it is attained. What toil and patience have gone before; what days and nights of watching and weariness; how often hope deferred made the heart sick; how year after year has dragged on, and seen the end still afar off—all that counts for little, if the long struggles do not close in victory. And yet in the history of human achievements, it is necessary to trace these beginnings step by step, if we learn the lesson they teach, that it is only out of heroic patience and perseverance that anything truly great is born."

— H. M. Field, 1866

Hibernia has been a long time in the making—actually much longer than most people realize. We remember the various announcements, the political wrangling, the construction milestones and setbacks, forgetting the events dating back millions of years that provided the opportunity and challenge of harvesting a precious natural resource from far beneath the cold Atlantic Ocean off the east coast of Newfoundland. Those who know Newfoundland's frequently fogbound coasts and biting winds may find it difficult to believe that 100-150 million years ago Canada's east coast was as hot and humid as the Caribbean. At that time, the area 315 kilometres east southeast of St. John's, where the Hibernia oilfield is now located on the Grand Banks, was a series of river channel systems close to the shoreline. These have since been buried under 3000 metres of mud, sand, and organic material including the remains of marine plants and life forms of all sizes. Over millions of years, layer after layer of this sediment was added to the ocean basin, the weight of the enormous build-up turning it into shale, sandstone, and other types of rock. And deep within the basin, the nature-prescribed combination of tremendous pressure, high temperature, and bacterial action caused the conversion of organic material into natural oil. Nestled in porous sandstone, it has been trapped for centuries after a series of folding and faulting by a layer of impermeable shale.

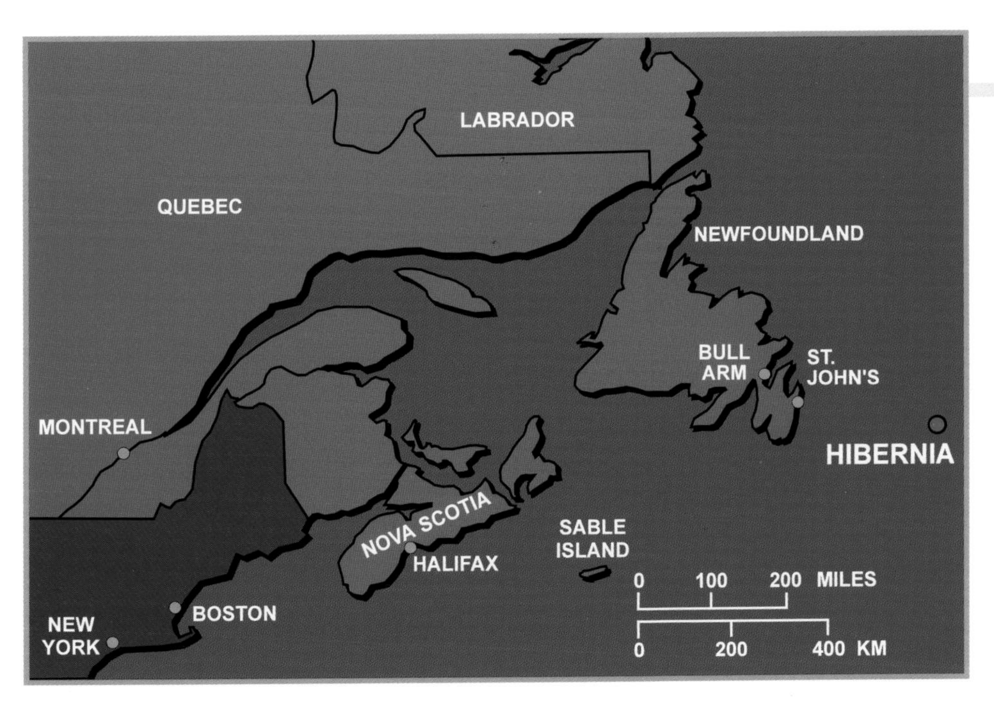

More than 100 million years ago, what is now known as the Hibernia area was part of a supercontinent. When this enormous land mass was pulled apart by continental drift the Atlantic Ocean was created, along with the faults that eventually became the Hibernia field.

There it remained undisturbed until the Hibernia sandstones, included in acreage owned by Mobil Oil Canada, were penetrated by the drillship *Glomar Atlantic* under the supervision of Chevron Canada Resources in 1979. The discovery was one of the most important developments made in the last century by the Canadian oil industry. This exploratory drilling venture had been prompted by promising seismic surveys. Sound waves echoing off the ocean floor indicated that the geological formations under the Continental Shelf off Newfoundland were capable of containing fossil fuels. They certainly were, and further drilling confirmed the presence of a giant oilfield holding an estimated three billion barrels of high-quality crude oil in its Hibernia and Avalon Reservoirs, 615 million of which are estimated to be recoverable.

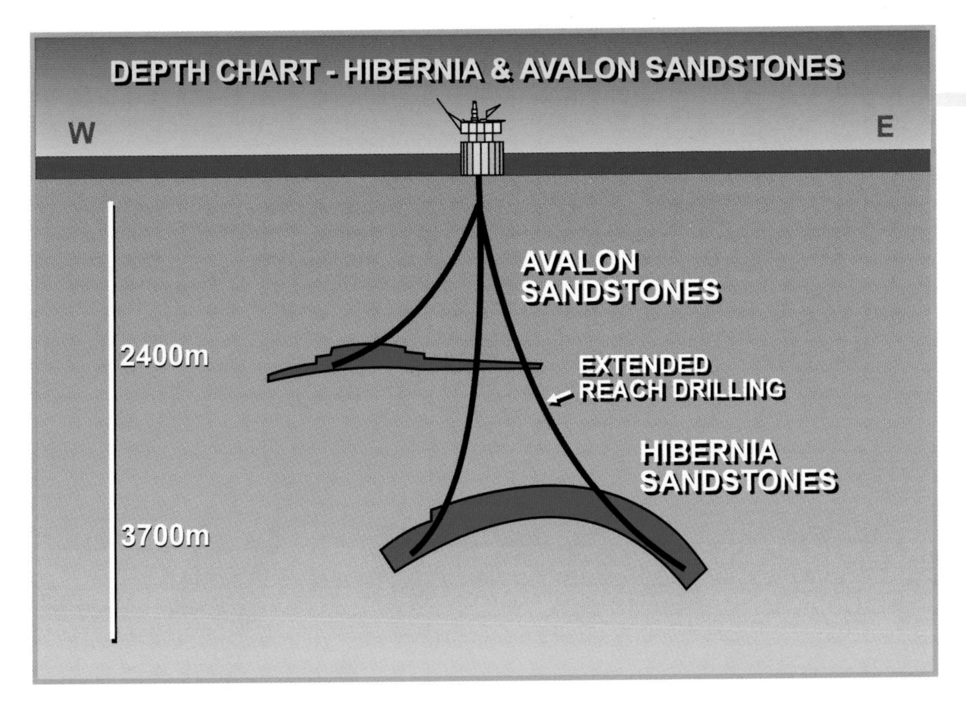

The two productive oil reservoirs in the Hibernia Field are the Hibernia Sandstones at a depth of 3700 metres and the Avalon Sandstones at 2400 metres.

Breakrite
ODC
BRICKLAYERS &
ALLIED CRAFTSMEN

A RIGGS REMINDER
FALLS KILL
Be Sure
to tie off!
HIBERNIA – 1995

LIEBHERR
Coleman

The first oil exploration permits for the Grand Banks area had been granted to Mobil Oil by the federal government during Joey Smallwood's term as premier of Newfoundland and Labrador. In 1964 Smallwood had a bronze plaque placed on the floor of the Atlantic Ocean near the Virgin Rocks, a gesture that was meant to send the message that anything in the Grand Banks area—including petroleum—belonged to the province. With the 1979 discovery, Newfoundlanders and Labradorians learned that the Grand Banks, historically renowned for rich fishing grounds, did indeed have the necessary elements for producing oil: ancient rocks and ocean.

OUR HISTORY

On June 23, 1964, an expedition sponsored by the Government of Newfoundland and Labrador explored the Virgin Rocks area of the Grand Banks. Divers Hugh Lilly, Cal Trickett, and John Snow descended from the support vessel **Bamasteer**, and fastened a plaque to the ocean floor at a depth of 19 metres. It was the first time that people walked on the surface of the Grand Banks.

Courtesy of *The Evening Telegram*

The signing of the Hibernia Binding Agreement on September 14, 1990 was an historic time for Newfoundland and Labrador in more ways than one. For the first time, all five of the province's premiers to that date were together in one room. Pictured left to right: Tom Rideout, Frank D. Moores, Joseph R. Smallwood, Clyde K. Wells, and A. Brian Peckford.

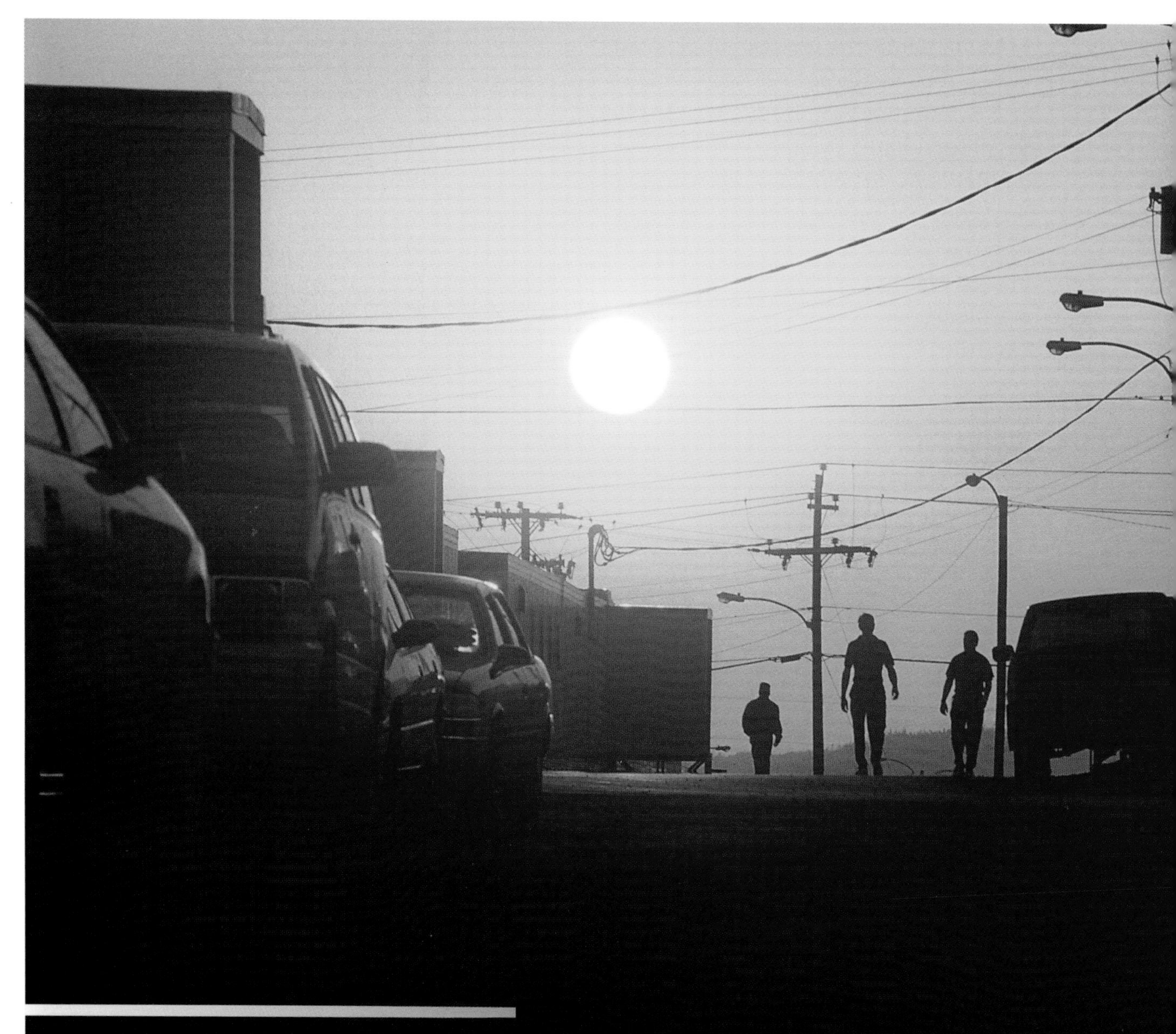

Our history has shaped us over the years
Through a mixture of hope, desperation and fears
But now a new fantasy beckons and calls
And our music is playing the Hibernia Waltz.

— *"Hibernia Waltz"* by Bud Davidge, recorded on Simani's *Saltwater Cowboys* album. Copyright 1981.

Since the federal government announced a moratorium on cod fishing off the coast of Newfoundland in 1992, many of the Newfoundlanders who had relied on the fishing industry have had to consider alternative ways to make their livings, to diversify and strengthen the provincial economy. Some have turned to the tourism industry, promoting the province's heritage, natural beauty, and other attractions. Many still look to the Atlantic, not only for alternative fisheries, but now in hopes that Hibernia will be a successful catalyst for the development of an oil industry in the province. They look to this new promise held by rock and sea, perhaps with thoughts reminiscent of former Premier Brian Peckford's words on the occasion of the signing of the Statement of Principles for the Development of the Hibernia Oilfield on July 18, 1988: "One day the sun will shine and have-not will be no more."

"This is one effort that will be rewarded in many ways for many years to come."

— Hon. Sandra Kelly, Tourism Minister, May 1997

PKS

REMINDER
1208
YES

"I can only hope that things unfold in the future that we build more of these because the skill and ability is here. All we need now is the opportunity."

— Clyde Wells, then Premier of Newfoundland and Labrador, on the occasion of the roll out of the completed M-20 module from the fabrication hall to the assembly pier

Successive premiers have been supportive of the oil development, but more cautious than Peckford, known for his flair for the dramatic. "In spite of the temptation to become infatuated with Hibernia and its many opportunities, our strategy must be to continue to pursue a broad range of non-oil related endeavors," said then Premier Clyde Wells on September 14, 1990, when the Binding Agreement was signed. Residents of the province have become more cautious in their optimism, checking expectations that might have been too high. "The expectations will always be too high for a have-not province like ours," said John Crosbie at the 11th International Offshore Petroleum Conference in 1995. "Who can blame us for being optimists living in our sometimes harsh environment, with our history and knowing as we do what the suffering being the sport of historic misfortune means."

"If we stay in construction for thirty years we might never see anything like Hibernia again."

— Juan Beckett of Gambo with Robert Goodyear of Ladle Cove, Rodmen

Since the time of the discovery, advocates who view the Hibernia Project as a possible starting point for a new industry have also met with incredulity, cynicism, or at least reserved hope, tempered by the memories of other ventures which failed to deliver. The topic of the Hibernia Project has spurred intense debate in local and national media, in political circles, and around kitchen tables. It has been criticized as an expensive "make-work" project and an imprudent gamble with taxpayers' dollars. In spite of opposition and setbacks, Hibernia has proceeded in an impressive marriage of perseverance and skill. The eyes of the many workers and other supporters who have invested time, effort, and expertise in the oil development are on the future.

The *Garden City* ferried GBS workers to and from the deepwater construction site.

Ken Hull, who assumed the presidency of the Hibernia Management and Development Company Ltd. (HMDC) in June of 1994 and held the position until September of 1996, shared his thoughts on looking to the future with those in attendance at the 1995 International Offshore Petroleum Conference hosted by Newfoundland Ocean Industries Association in St. John's. "I grew up in Saskatchewan, where one of the rites of passage was learning how to drive a tractor," he said, and related how his first day driving a tractor was "a disaster," as he cultivated crooked rows in the field. Hull's father surveyed the situation and offered some advice: "You can't cultivate a straight row if you continually look behind. You must keep looking ahead—and always remember where you have been." "That lesson," Hull told his audience, "has been as helpful driving corporate change as it has been driving that tractor. Keep focused on the challenges of the future, and always remember the values and traditions that proved successful in the past."

"High productivity, a commitment to a team environment, cost effective operations, and positive attitudes will cement a place for Newfoundland in this new industry. And as always, Newfoundland's strengths will lie in the ingenuity and persistence of its business and people."

— Ken Hull, then President of Hibernia Management and Development Company Ltd., June 1995

Among Hibernia's strongest supporters, who can certainly appreciate Hull's philosophy, is John Crosbie, described by fisheries aficionado Cabot Martin as a bulldog that wouldn't let go because of his determination to see the Hibernia Project become a reality for the people of Newfoundland and Labrador. Crosbie's dedication epitomizes the spirit of what Hibernia has become, and he has been quick to point out in thanks and congratulations the contributions of many individuals to realizing the opportunity that Hibernia presents.
In his position as Federal Minister of Fisheries and Oceans, Crosbie performed the unenviable task of making the public announcement of the moratorium on cod fishing. In more recent years he has been able to take a more inspiring role, attempting to prod industry and political leaders and the public into capitalizing on offshore opportunities. At the 11th International Offshore Petroleum Conference in 1995, Crosbie maintained, "It is up to us as Newfoundlanders to show that we did not waste the opportunities provided for us. As Sophocles said in 408 B.C., 'Opportunity has power over all things.' Our opportunity is Hibernia."

Crane Operator Gerald Glode on a GBS tower crane.

UNITED ASSOCIATION
NEWFOUNDLAND LABRADOR
LAKIN

Members of the International Brotherhood of Electrical Workers (local 2330) at the Hibernia construction site collected nearly $1000 in just two days during the 1995 Christmas season to assist a family in need.

MANITOWOC
BOILERMAKERS

Four topsides mounted structures were fabricated at Bull Arm: helideck, flare boom, main lifeboat station, and auxiliary lifeboat station. Here the flare boom is being lifted onto the module.

"We've had some struggles, but have succeeded."

— Janet Lahey, Administrative Assistant

Seizing such opportunities is apt to be risky business, so that ventures like Hibernia epitomize the adage "Nothing ventured, nothing gained." From the first instances of oil exploration off the east coast in the 1960's until Hibernia oil hits the market, risk is a factor of which all those involved in the inception and development of the project cannot help but be keenly aware. Two hundred and fifty exploratory wells have been drilled off Canada's east coast in the last several decades, sometimes yielding little more than disappointment. In fact, statistics on exploratory drilling indicate that, even with reliable seismic information, only one drill hole out of 10 will encounter success. But all great enterprises counter risk with dedicated effort and skill, as many undertakings that have left their mark on Newfoundland and the world attest. In fact, Hibernia takes its name from one such historical success.

Aboard the S.S. Hibernia, 1873

Scale of Provisions to be allowed and served out to the Crew during the voyage, in addition to the daily issue of Lime and Lemon Juice and sugar, or other anti-scorbutics in any case required by Law:

Sunday, Tuesday and Thursday:
one-half lb. beef, one-half lb. flour

Monday, Wednesday, and Friday:
one-half lb. pork, one-third pint peas

Saturday:
one-half lb. beef

daily:	**weekly**:
3 quarts water	14 oz. sugar
one-half oz. coffee	one-half lb. rice
one-eighth oz. tea	
1 lb. bread	

Source: Maritime History Archive, *Hibernia* Agreement and Account of Crew, 1873.

Hibernia

Food Facts

The monthly consumption of food for the Hibernia cafeteria at Bull Arm was somewhat higher than the average household.

How would you like to take this grocery list to your local supermarket?

Monthly Shopping List

13,000 ...... dozen eggs
16,000 ...... pounds of potatoes
19,000 ...... steaks
28,000 ...... loaves of bread
30,000 ...... litres fresh milk
2500....... cases fruit

Did you know?

More than 120 food deliveries were made to the cafeteria each month.

There were six 100-gallon soup pots in the kitchen.

The dining hall of the cafeteria could seat 1000 people at a time.

Hibernia
JACKIE
BH1

The steamship *Hibernia* was built at Jarrow, England in 1861. At around 110 metres long and 2164 net tonnes, it would today appear dwarfed next to the 1.2 million tonne Hibernia platform, but during its day the *S.S. Hibernia* was very capable of performing its duties of laying and repairing telegraph cables in oceans around the world, including Newfoundland. Here it assisted the *Great Eastern*, the largest steamship in the world at the time, in laying the shore ends of cables stretching from Valentia, Ireland to Heart's Content, Newfoundland in 1873 and 1874. The first mission was completed on July 5th, 1873 and declared "cause for universal congratulation" by the *North Star* newspaper. In August of 1874, in a successful repeat of the previous year's accomplishment, the *Hibernia* landed another cable in front of the Heart's Content Cable Station.

Far top: "Steam engine and lifting gear on the *Great Eastern*, 1866."

— MF-126, Donard de Cogan Collection, Maritime History Archive, Memorial University of Newfoundland.

Above: "Paying out gear on the *Great Eastern*, 1866."

— Collection of The Institution of Electrical Engineers Archives, London, England.

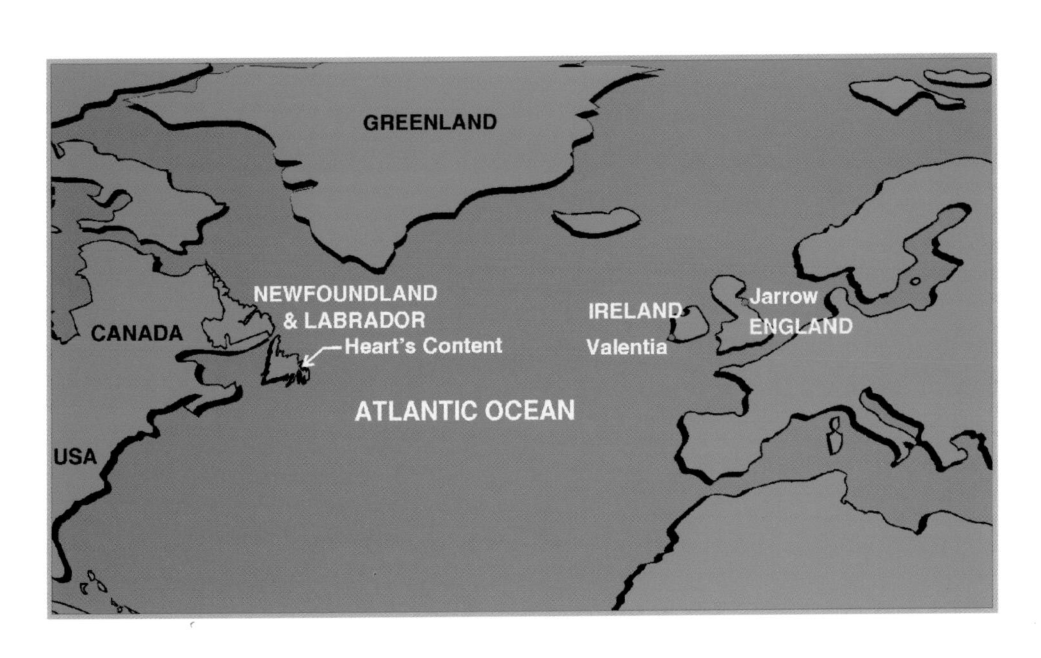

Newfoundland had come to the forefront of telegraph communications in 1866 when a telegraph cable was successfully laid across the ocean from Valentia to Heart's Content. But it was not a victory easily won. The first attempts at laying a trans-Atlantic cable had been in 1857 and 1858. In a wonderful historical coincidence, the North American ends of these attempts were concentrated at what was then known as Bay of Bull's Arm, and became the Hibernia construction site, Bull Arm, in 1990. A nineteenth century correspondent described the area in a letter:

> *All who have visited Trinity Bay, Newfoundland, with one consent allow it to be one of the most beautiful sheets of water they have ever set eyes upon. Its color is very peculiar—an inexpressible mingling of the pure blue ocean with the deep evergreen woodlands and the serene blue sky. Its extreme length is about eighty miles, its breadth about thirty miles, opening boldly into the Atlantic on the northern side of the island. At its south-western shore it branches into the Bay of Bull's Arm, which is a quiet, safe, and beautiful harbour, about two miles in breadth, and nine or ten in length, running in a direction north-west.*
>
> *The depth of water is sufficient for the largest vessels. The tide rises seven or eight feet, and the bay terminates in a beautiful sand-beach. The shore is clothed with dark green fir-trees, which, mixed with birch and mountain-ash, present a pleasing contrast. The land rises gradually from the water all around, so as to afford one of the most agreeable town sites in all the island.*
>
> — H. M. Field's *History of the Atlantic Telegraph*, 1866

Despite these apparently favourable conditions, Bull Arm never became a long-term telegraph station site. The 1857 cable broke before reaching the shore at Bull Arm, and the 1858 cable ceased to transmit messages after only one month.

Still American financier Cyrus Field, who spearheaded these attempts, persisted, rallying for investors' support and seeing to the necessary improvements in design and operation. Yet another attempt in 1865 failed, but on July 27, 1866 the *Great Eastern* and its accompanying ships, the *H.M.S. Albany* and *Medway*, successfully completed the trans-Atlantic laying of a telegraph cable at the new destination of Heart's Content, where a cable station would remain in operation until 1965. Upon the announcement of this success, royalty and political leaders rushed messages of congratulations to Cyrus Field and his party, and the *London Times* declared the cable "the glory of the age and nation." Amid the excitement of accomplishment, Field's brother, Henry, recalled the struggle that led to this victory, the patience, toil, and perseverance invested and the ridicule endured. This compelled him to open his book, *History of the Atlantic Telegraph*, with the stinging observation "Mankind worship success but think too little of the means by which it is attained."

Bull Arm, Trinity Bay, Newfoundland. Exterior view of Telegraph House in 1857-1858. Lithograph by R.M. Bryson from a drawing by R. Dudley. Collection of the Newfoundland Museum.

"I'm proud to be a Newfoundlander and a Bull Arm worker."

— Eric Simmons, Site Security Chief, 1996

NO
SMOKING
Major Offshore
BAKE
SHOP

"I hope that when the first barrel of oil is produced from Hibernia, we will be able to look back with a deep sense of pride at a job well done."

— Hon. Rex Gibbons, then Minister of Mines and Energy, September 14, 1990, the occasion of the signing of the Hibernia Development Agreement

The story of the trans-Atlantic cable is one which the Hibernia Project has paralleled in many ways: searching for financial backing, facing public incredulity, and disheartening setbacks have been countered by enthusiasm, determination, technological innovation, and progress. Now we witness history further repeating itself, as Hibernia enjoys the same celebrated final success as the historic pioneering venture in which its namesake participated in more than a century ago.

Hibernia's company vision, "People Pioneering Offshore Excellence," acknowledges the novelty of the project in Newfoundland. It is a trailblazing effort involving technology transfer and experience building, aiming to establish a place for Canada and Newfoundland at the forefront of the oil industry. Derm Cain, President of the Oil Development Council, which represented the unions at the Bull Arm construction site, said in the winter of 1991, "We aren't building a concrete platform. We think we're building a new industry. This is the first time this has been built in North America. If we're successful—and we're going to be successful—I think it's the birth of a new industry in this province."

Many workers and staff at Bull Arm shared Cain's optimistic foresight, including Information Systems Site Coordinator Dan Tuttle of Mount Pearl, who spent more than five years working on the project and said, "It's going well now, and hopefully it's been a kind of starting block for another. That's the big thing—a stepping stone for more."

"Early development in the North Sea was as tough or tougher to get started, but once the first project was producing, there was always another. Hibernia is not an isolated project — it's the beginning of an offshore industry for Newfoundland."

— Christine Fagan, then President of Newfoundland Ocean Industries Association, 1992

"There's virtually nobody here who has worked on a bigger project than this. I've worked on a lot of jobs in different places, but nothing comes close to it."

— David Sverre, Assistant GBS Construction Manager at Bull Arm

"It's such a big project that it's something you can remember and look back on and take with you pretty much anywhere you go."

— Dan Tuttle, Information Systems Site Coordinator at Bull Arm

Steve Bragg of St. John's was the Bull Arm site's HEART (Hibernia Employees Achieving Results Together) Program Coordinator, arranging for the presentation of safety awards to deserving workers, as well as awards for good cost control, safety, or productivity suggestions. He is a former instructor of the Marine Institute in St. John's and sees the Hibernia enterprise as a history-making development for Newfoundland. "I've always wanted to get involved with the Hibernia Project even back during my teaching days," he says, "I always said, 'If that happens, I'm going to be there.' So I actively tried to get involved in the project. I see it as the birth of a new industry for this province."

A Developing Industry
Kevin Tobin's rendering of Premier Brian Tobin as an oil tycoon capitalizing on the 406-million barrels of recoverable Terra Nova oil alludes to the song "Saltwater Cowboys," made popular by the duo Simani, and to the late Premier and Father of Confederation Joey Smallwood and his days as radio's Barrelman.

NCN
PKS
PROJECT
HIBERNIA - 1995
CIVIL CONSTRUCTION
"I Slipped on the GBS!"
NC
IRON WORKERS
LOCAL 764
FRED SMITH

Indeed, arrangements for the development of Terra Nova, the second largest discovered oil field off the east coast of Canada, are under way. When Newfoundland and the companies committed to the development of the Terra Nova Oil Field signed a letter of intent on August 5, 1996, Premier Brian Tobin declared: "This announcement marks a major milestone in the development of the province's emerging oil industry. Building on our experience in the construction of the Hibernia platform, the province is now poised to benefit from another large oil field development."

As well as being a pioneer in the province's oil industry, Hibernia is venturing into virgin territory with technological innovations. The Gravity Base Structure (GBS) design chosen in 1985 has proven itself in the North Sea, and so is considered a worthy design for the similar Grand Banks environment. A continuous process of placing rebar and pouring cement called slipforming was used to construct the Hibernia GBS, which is comprised of 450,000 tonnes of concrete and 100,000 tonnes of reinforcing steel. Bull Arm workers laboured around the clock and in winter storms, making the implementation of this kind of concrete technology a success.

Mixing Concrete: The method of pouring concrete for the GBS was changed from jumpforming to slipforming in 1994. This efficient method helped achieve every subsequent GBS milestone on schedule and within budget, with construction complete on November 1, 1996.

The GBS was partially constructed in a drydock at Bull Arm and then floated to its deepwater site. Before this event actually happened, many visitors and workers alike expressed incredulity about the procedure. Gloria Warren-Slade, Tour Coordinator at the Bull Arm Construction Site, recalls how one woman visiting from England listened to her explain how the GBS would be moved from the drydock, then approached her, saying, "You mean you're going to float concrete?.... Clever."

Essentially, floating the massive concrete structure involved taking advantage of what is known as Archimedes Principle. According to this principle, an object that is placed in liquid experiences an upward push (buoyancy) equal to the weight of the fluid it displaces. Even though the GBS is a gigantic mass of concrete and rebar, it floated once the drydock was flooded because the volume of water it displaces is greater than its own weight.

Gloria Warren-Slade spent four years organizing and guiding tours of the Hibernia Construction Site in Bull Arm, during which time over 65,000 people visited. "I showed 96 12-year-olds around the site on my last official tour day," she said. "So it was a happy, sad, crazy, and exciting occasion all at once."

Gloria Warren Slade
Public Affairs
HIBERNIA
GLORIA WARREN

Hibernia
GBS Batch Plant

HMDC's early corporate slogan "Building on a Solid Base" was fitting for the construction of the GBS, comprised of 450,000 tonnes of concrete and 100,000 tonnes of rebar.

ALEX
Garden
City
Garden
City

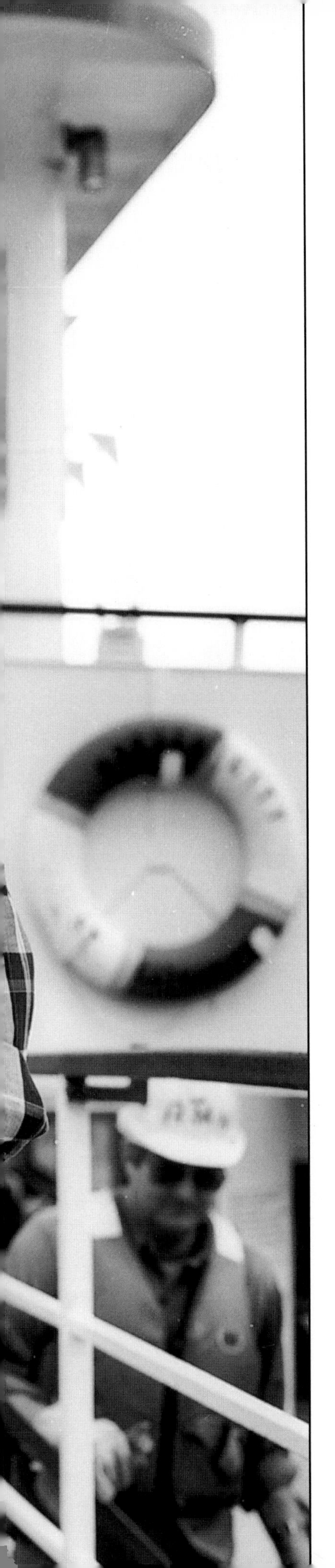

"The Oil Development Allied Trades Council sees the hook-up, mating, and tow-out of the Hibernia platform as one of the most significant events in the history of this province."

— Derm Cain, ODC President, September 1996

Interior view of the GBS serrated ice wall.

Hibernia's engineers have supplemented the reliable 106-metre GBS with a tailor-designed feature that makes it the first of its kind: a concrete ice wall serrated with 16 teeth to withstand possible collision with icebergs, which are carried to the Grand Banks from Greenland and Ellesmere Island by the cold Labrador Current flowing along the Continental Shelf. Although very large icebergs cannot drift into the area of the Hibernia Oilfield because of relatively shallow water, engineers recognize the damage that even smaller icebergs can do. So they have erred on the side of safety, designing the ice wall to protect the more than one million barrels of crude oil that will be kept in the inner storage cells of the GBS from an impact of up to six million tonnes.

Building the Gravity Base Structure

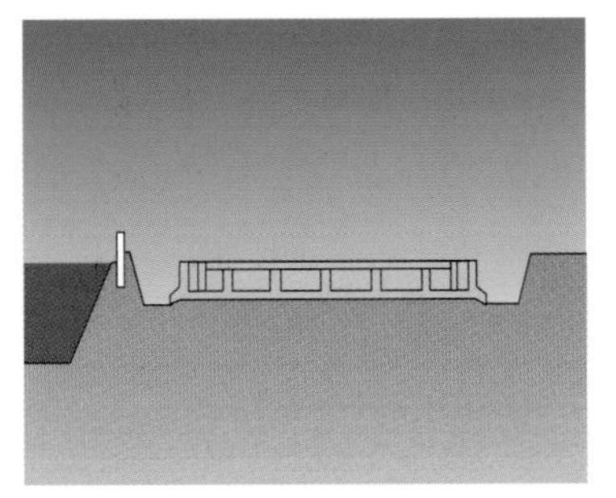

1. Construction of Base Slab in drydock

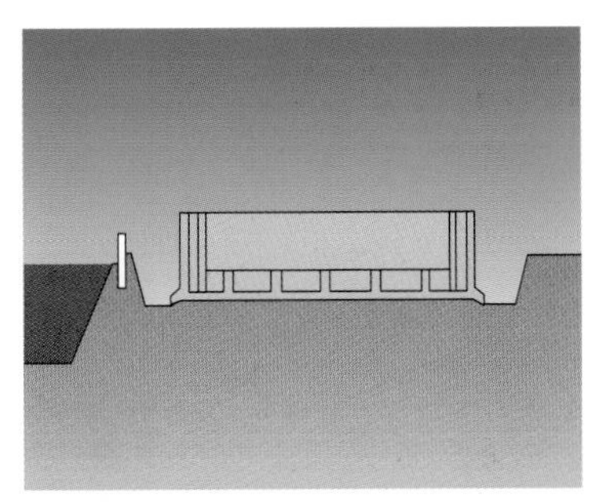

2. Construction of Lower Caisson Walls

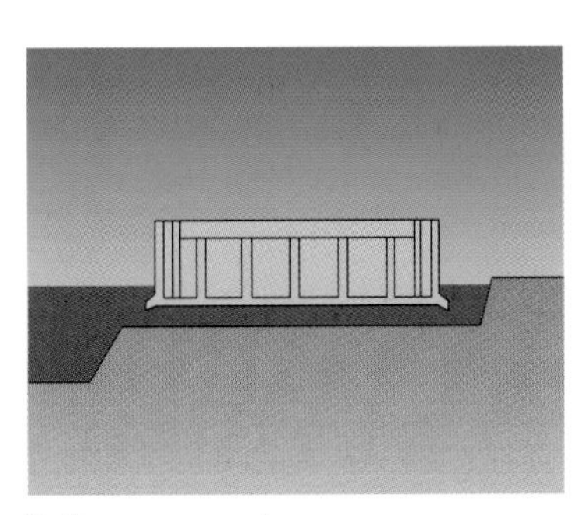

3. Berm removal

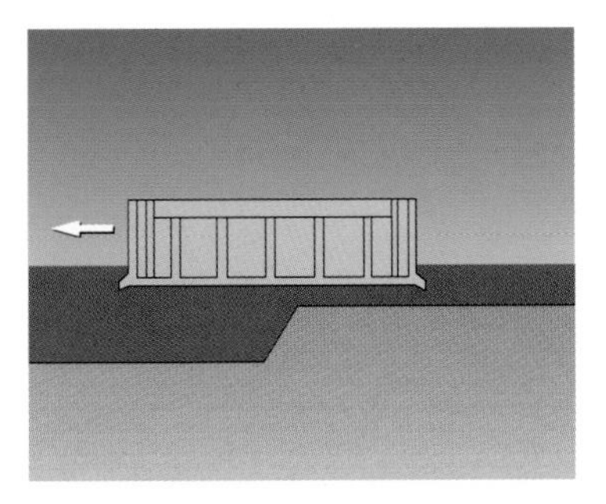

4. Float out from drydock

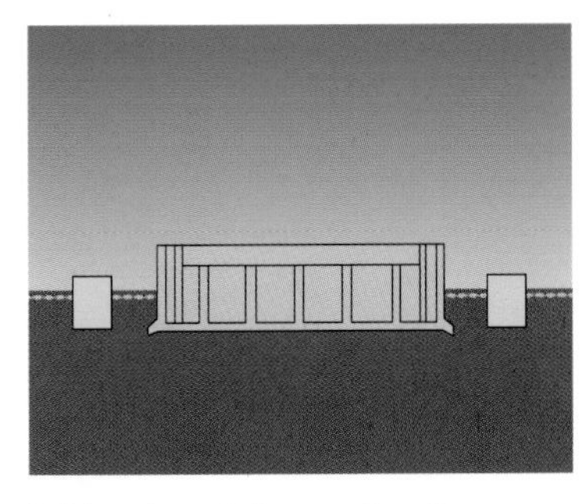

5. Mooring at deepwater site

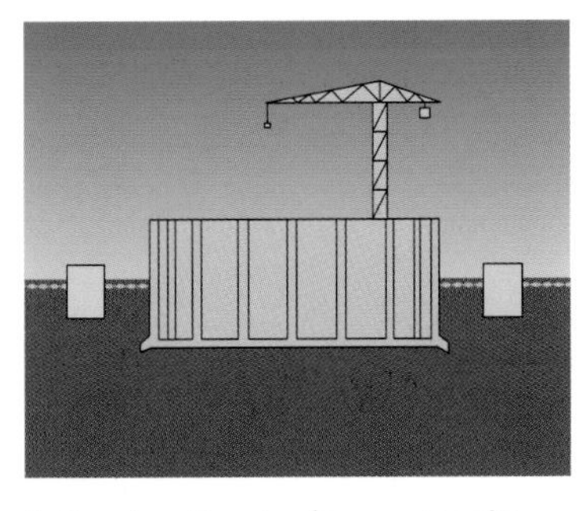

6. Construction in deepwater site

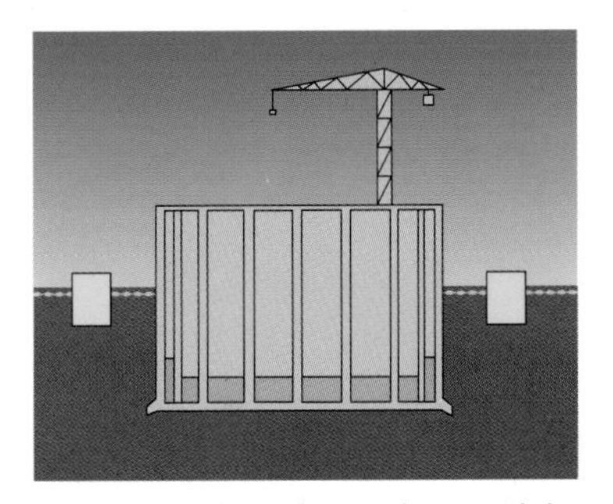

7. Construction of Top Closure Slab

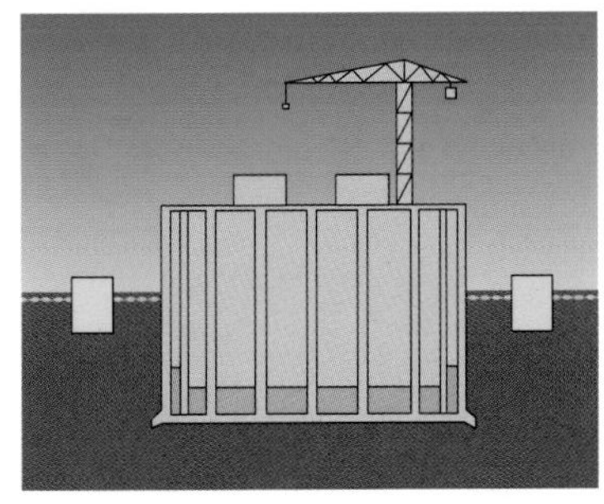

8. Construction of shafts

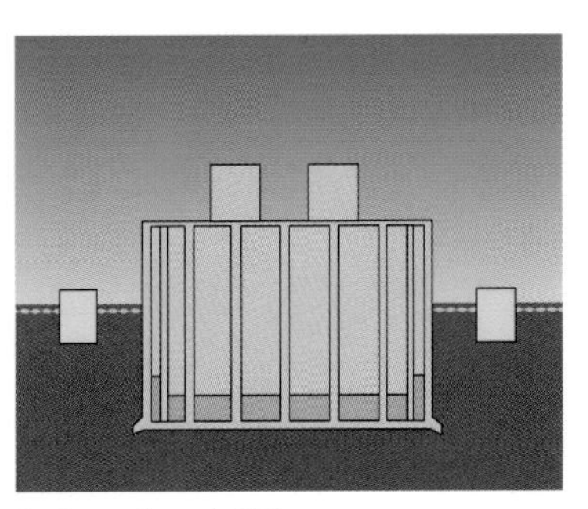

9. Completed GBS

Another first of its kind innovation, a floating batch plant, supplied the 450,000 tonnes of super high-strength concrete to the GBS deepwater construction site. Even Hibernia's helideck, installed in August of 1995 on the M-50 Living Quarters/Services Module, boasts a special high-friction mineral-laden coating intended to keep workers and equipment firmly and safely grounded when the wind blows high. This will be its first use in North America.

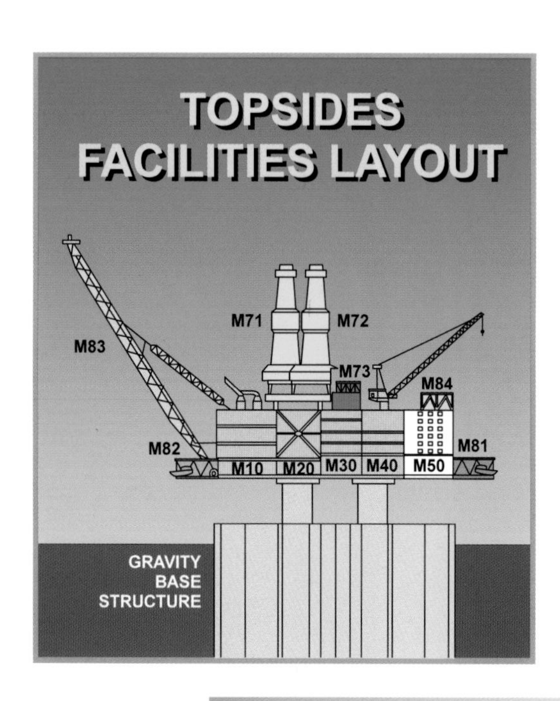

M10 - Process Module
M20 - Wellhead Module
M30 - Mud Module
M40 - Utility Module
M50 - Living Quarters/Services Module
M71 - East Derrick
M72 - West Derrick
M73 - Piperack Structure
M81 - Main Lifeboat Station
M82 - Auxiliary Lifeboat Station
M83 - Flare Boom
M84 - Helideck

Hibernia

"Hibernia is neither a make-work project nor a get-rich scheme. It's somewhere in the middle. It's had a long and controversial history, but it's now looking pretty positive."

— James Feehan, Economics professor at Memorial University of Newfoundland, May 1997

LOCAL 720
STEVE REID
PIPE SHOP
AM

The icewall and helideck coating are just two examples of the platform's safety-conscious designing. Government regulations require that Hibernia have two evacuation systems: lifeboats and escape chutes. Hibernia has actually exceeded that requirement. "Our commitment to safety compelled us to use three evacuation systems on the platform," says Paul Kent of the Environment, Loss Prevention and Quality team. This third option, the GEMEVAC system, allows for the total evacuation of the platform in about 90 minutes. It is a dry evacuation system, so that evacuees are transported from the platform in an enclosed gondola onto the two Maersk platform support vessels.

WHAT'S IN A NAME?

In a contest conducted by HMDC, two Newfoundland students named the two Hibernia multi-function Maersk platform support vessels constructed at Marystown Shipyard. Tommy Walsh of St. Joseph's High School chose the name **Nascopie** to honour Labrador's aboriginal inhabitants. Scott Hayes of Christ the King School decided on **Norseman** "because they discovered our land and should get recognition."

Whether the lifeboat, escape chute, or GEMEVAC method is used would depend upon the specific situation. "We provide several options and rely on training, knowledge and the experience of the platform command team to decide which method is the most appropriate," says Kent. Every Newfoundlander who remembers the devastating winter storm of February 15, 1982 in which 84 men were lost when the semisubmersible oil rig Ocean Ranger capsized on the Grand Banks can appreciate such cautious, environmentally responsive designing, training, and planning.

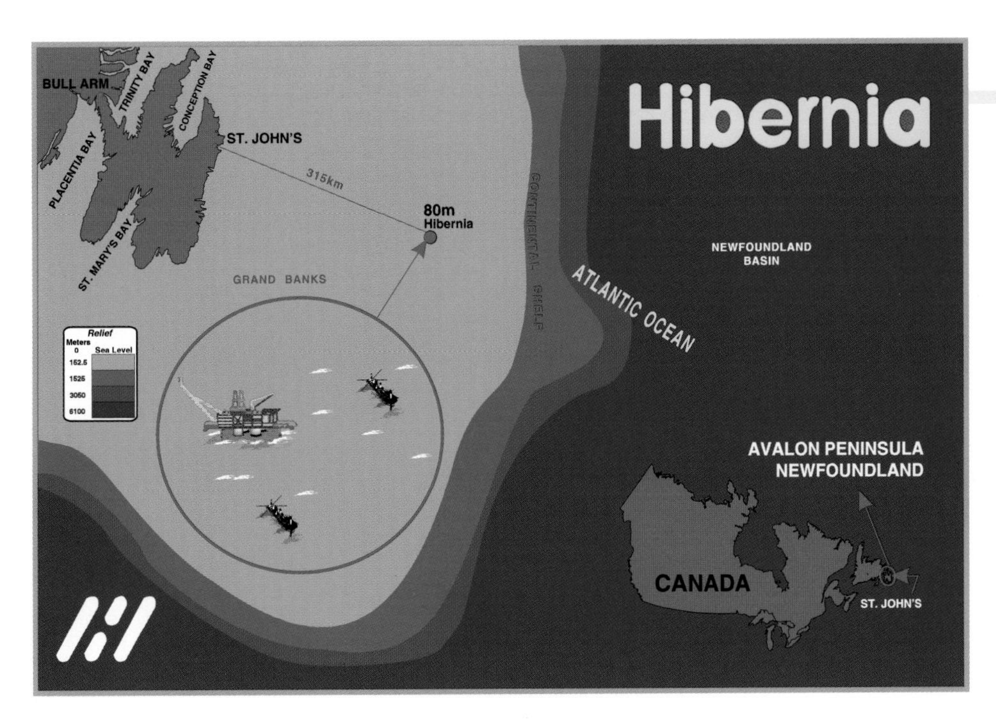

Twenty-five years of exploration and pre-development engineering on the Grand Banks of Newfoundland reached a climax with the Hibernia Binding Agreement in 1990, marking the beginning of the development and construction phases. It is estimated that 615 million barrels of oil can be recovered from the Hibernia oilfield located 315 kilometres east south-east of Newfoundland's capital city, St. John's, in 80 metres of water.

Brenda Wiseman and Vernice MacKenzie take advantage of berry patches at the Bull Arm Site.

"I'm just very impressed by the quality of the people here, not only from a technical standpoint, but from a human one as well."

— Sandy Cornelius, Well Performance Team Lead, February 1997

5456TS

During construction of the Bull Arm site, there was plenty of opportunity for innovation and learning for people involved with all facets of the project. Tony Hylton, Topsides Site Manager and a professional engineer originally from England and now a Canadian citizen, has worked on large construction projects in his native country, in the North Sea, and elsewhere in Canada. He has appreciated the challenge of being involved with a frontier project: "It's not many times that you get the opportunity to actually get involved with a facility that was nothing more than the forest and the pond and then develop the facility and build a module. You face all the problems that you have in doing that: in training a workforce, helping develop an infrastructure, and then receiving modules from overseas, and also being involved in the hook-up and commissioning. As an engineer you're going right through pretty well every engineering aspect that you could."

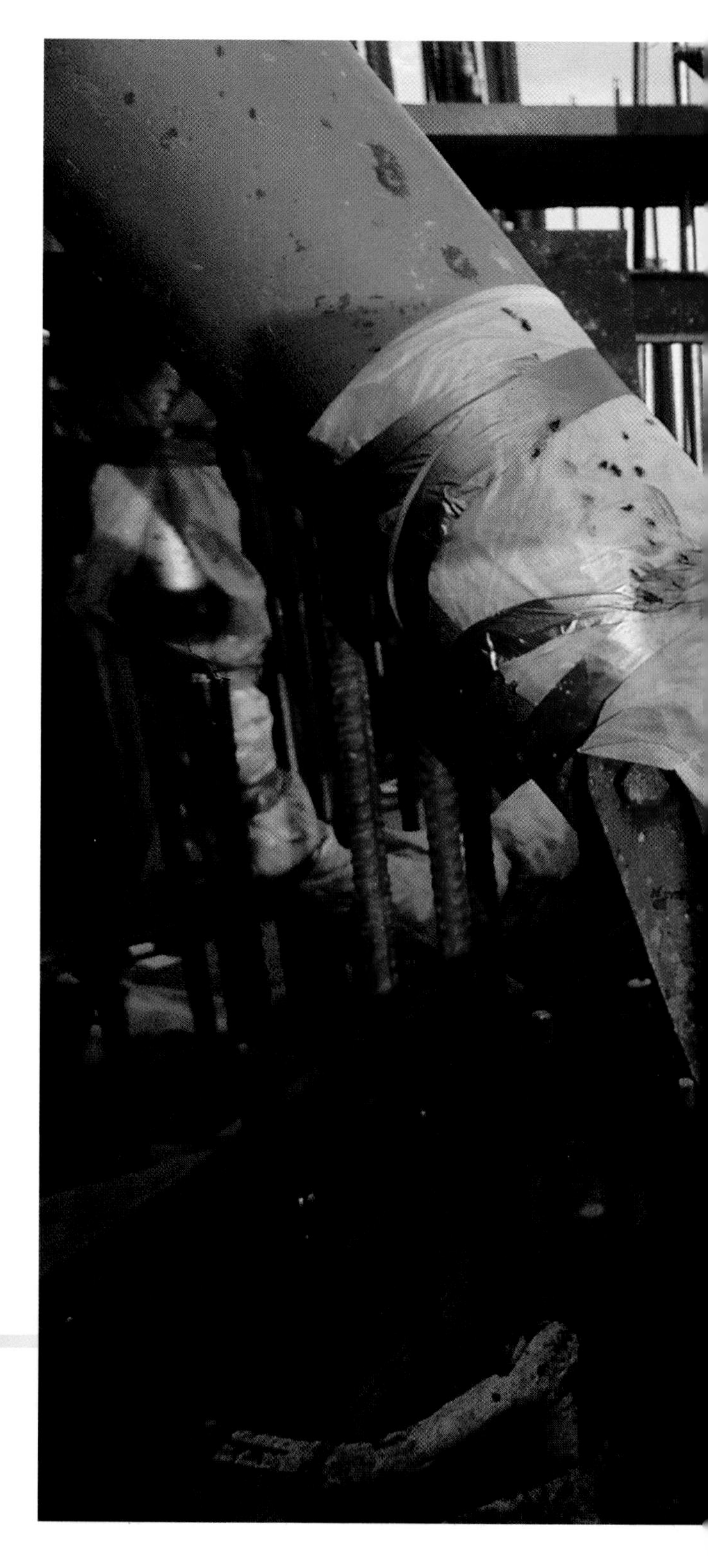

Gerry Boone, Cement Pourer on the GBS.

Heading Home: For some workers, employment at Bull Arm was their first experience with living in a construction camp. This often entails adjusting to shift work and extended periods of time away from home. For Gus Doyle of Keligrews, who began working with the first scaffolding crew at the site, the hardest part of camp life was being away from his wife and children. "We spend a lot of time talking on the phone," he said in 1993, "Aside from that, everything is pretty good."

The 4000-acre construction site eventually included all the amenities to keep a small town with a population of about 3400 people running smoothly: complete water and sewer; accommodation units; a post office, convenience store, bank, and hair salon; fitness and recreation facilities; a cafeteria with a seating capacity for 1000 hungry workers; a fire department, and medical facilities. As an elderly man taking a bus tour of the site on a warm Sunday afternoon in August of 1996 remarked, "It's all here, buddy!"

"You have to go beyond Hibernia and realize that there's a whole industry out there and we're just at the beginning."

— Duncan Mathieson, analyst at Gordon Capital Corp. of Toronto, May 1997

PASSB

JUST DO IT
NFLD. STYLE

On August 27, 1996 Arne Henriksen, President of Aker Stord Newfoundland, announced that the company would be hiring 60 Newfoundland welders, most of whom had worked on the Hibernia platform, to work at its Norwegian facility.

Welding foreman Bernie Wall has worked in construction camps throughout Newfoundland and Labrador, as well as outside the province. At Bull Arm he was responsible for welding operations at the Topsides Assembly Hall, and remembers his introduction to the site: "When I came here as first foreman on Topsides, the only thing here was a big pile of snow. We had to get a bulldozer to clear away the area to start steel erection." Wall witnessed much of the development of the site's facilities, as well as the first concrete pour for the GBS, and in 1993 he called Bull Arm "my home away from home."

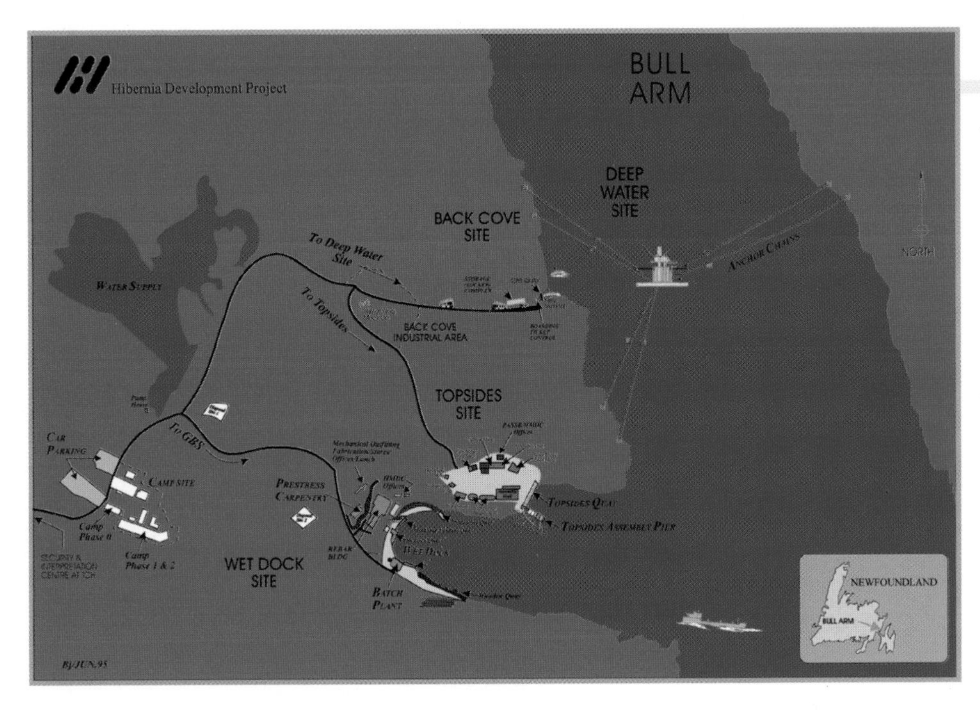

The Bull Arm, Trinity Bay construction site was chosen after various environmental and engineering evaluations identified it as a viable location with sufficient near-shore water depth. Further engineering work led to the layout of the site's construction and camp facilities, and environmental effects were continuously monitored over the construction period.

Tailoring various services to a large construction site requires careful planning and an array of adjustments throughout the phases of construction, and the personnel at Bull Arm adapted admirably. Ken Putt of Goulds, First Cook at the site's dining hall, has extensive experience in the restaurant industry and remembers how the kitchen and catering staff had to rethink food costs when it was realized that the average food serving on site would have to be approximately double that of most restaurants. He believes that such adjustments have contributed to his own learning experience: "I was used to working in restaurants that usually could seat maybe 80 to 90 people. Then I came out here into a dining room that can seat up to 1000 people at a time. So I must say I've gained a lot of experience here, a lot of knowledge that I feel I can move to another job and be able to take it over quite comfortably." Putt hopes that the Hibernia Project can do the same for its larger workforce. "Hibernia has put us in a good position to bid on projects all around the world," he says, "and I think that this is what you have to look at."

"Considering the state of the economy, I feel very privileged to have a good job here. I'm proud to be working here. And the fact that I don't have to go away to work like other Newfoundlanders makes me feel good about everything."

— Ken Putt, First Cook at the Bull Arm site

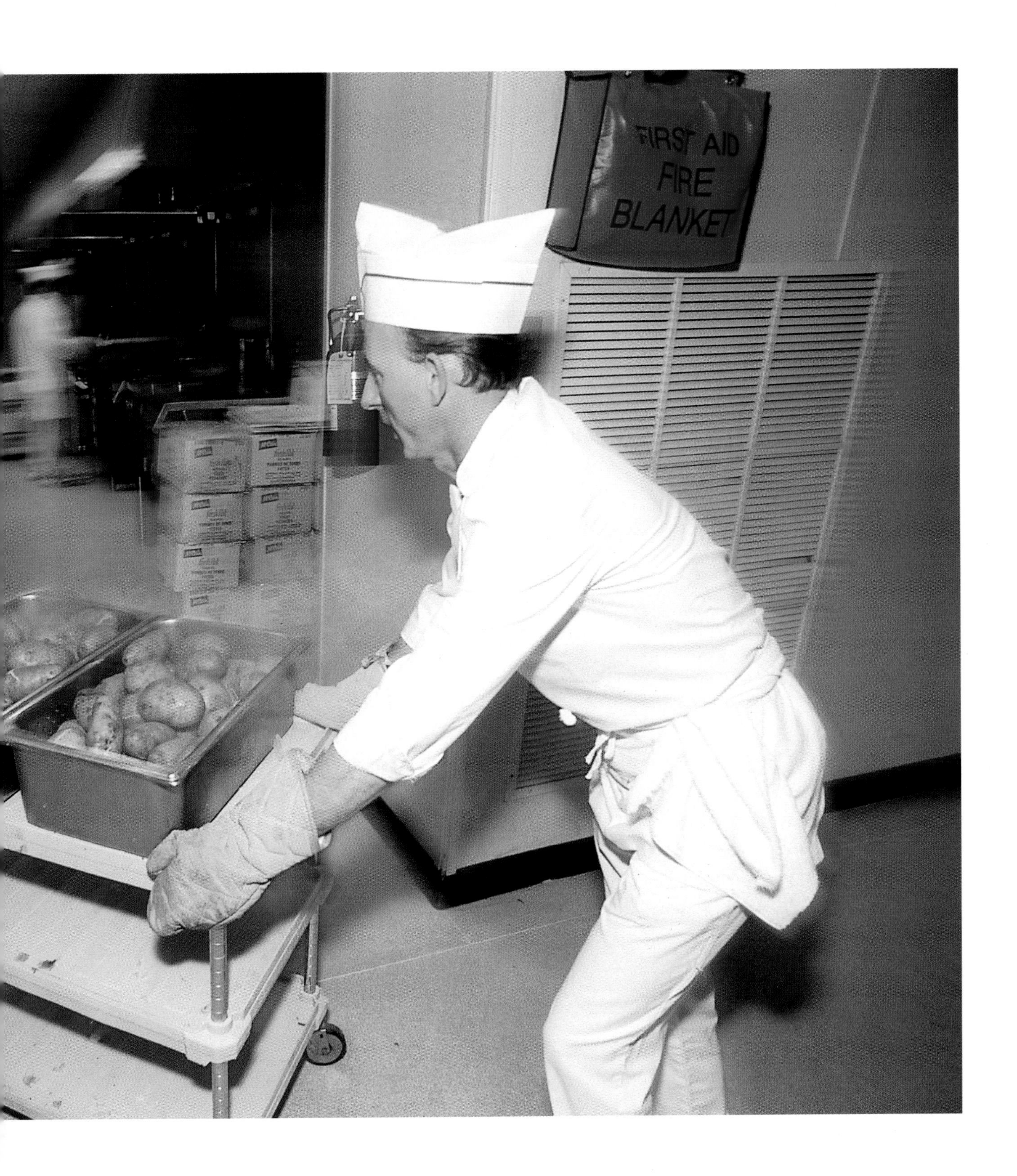
FIRST AID
FIRE
BLANKET

Mary Power, a Senior Occupational Health Nurse started work with Atlantic Offshore Medical Services at the Hibernia construction site in the summer of 1991 as part of a team providing emergency medical services and other health services to the site. She relays that she and her colleagues at Emergency Medical Services underwent a constant process of adjustment in reply to worksite changes. "The construction site changed, the demographics changed, the configuration of the workforce changed, the management changed."

Working at the site gave the EMS staff pre-hospital experience that they could not have gained anywhere else in the province. Like Putt, Power is enthusiastic about the transferable skills the job has given her, recognizing that expanded career options for people in her field are among the positive spin-offs of the project: "The pre-hospital care at Hibernia has afforded Newfoundlanders the opportunity to educate and practise the skills in industry, and we've made great milestones in industry with regards to pre-hospital care. We all feel a part of that.... As Newfoundlanders we'd like to see more of this calibre of a project come to Newfoundland, whether it's here or at Voisey's Bay or wherever. Certainly we feel proud of the service that we have provided, and really for the education and training that we've had the opportunity to get at Hibernia. We would like to see it really as the beginning of a new industry for emergency medical services."

"For us in medical services it's been interesting to work with so many people from all walks of life and to be able to provide them with a service. That in itself has been an experience."

—Mary Power, Senior Occupational Health Nurse at the Bull Arm Site

emergency

The 11 red overall-clad members of the Mosquito Cove Industrial Fire Department also monitored the health and safety of Hibernia workers. They are an amiable group who joke about themselves and each other and celebrate workmates' birthdays with good-natured pranks, but leave no doubt about their competency and qualifications. Each member has at least National Fire Prevention Association training and extensive experience in volunteer fire departments in their home communities or paid positions. While they are well-equipped to respond to fires and other emergencies, these men qualify their goals and responsibilities: they are not just firefighters, but Fire Prevention Officers who dedicated most of their time to enforcing safety regulations, doing inspections, and providing orientation training for new workers, as well as training the site's volunteer firefighters. Sometimes the officers' strictness in enforcing regulations put them out of favour with other workers, but department member Paul Noseworthy of Spaniard's Bay maintains, "If that saves one person's life, then it's all worth it." Most Hibernia workers used construction milestones to gauge their progress, but the Mosquito Cove Fire Prevention Officers pride themselves on things that never materialized. It is difficult to measure the contribution of preventative efforts, as the department's 'rookie,' Paul Snow, former chief of the Harbour Grace Volunteer Fire Department, realizes: "It's pretty hard to put a quantitative number on what didn't happen."

MOSQUITO COVE
FIRE DEPARTMENT
MOSQUITO COVE
FIRE DEPARTMENT

Because of unavailability of work in his field, Paul Snow had been unemployed before starting work for NODECO, one of the contractors at the Bull Arm site, and was more than eager to take on the position's responsibilities and long shifts. The job, he says, was "a godsend." Several of Snow's coworkers have similar stories. Harry Hiscock lives in Sunnyside, where he was a fire department member for over 15 years and chief for a portion of that time, and had been a long-time employee of the Long Harbour Phosphorous Plant in Placentia Bay, which closed in 1989 after more than 20 years of operation. He joined the Mosquito Cove Fire Department in October of 1992. "Long Harbour was closed down and I had been back there for three summers, and I got a call to come here that same week that I got my lay-off in there. So basically it was a godsend for me too," he says, "I was just ready to be unemployed when suddenly...I left there on Friday evening and came here on Monday morning. Timing was perfect."

COVE
DEPARTMENT

While there are many seasoned offshore construction industry professionals involved with the Hibernia Project, many people like members of the Mosquito Cove Industrial Fire Department were new to large construction sites and learned to apply knowledge and experience gleaned in other environments to their work at Bull Arm. The transition held surprises for some. John Gillis, originally from Cape Breton and living in Newfoundland for 23 years, is an employee of Metropol Security Company. He began work at Bull Arm in October of 1991: "I looked at it as a chance to participate in a bit of history. This is the first project of this type for both the province and our country—actually in North America." So Gillis retired early from the RCMP to accept a position as Site Security Manager at Bull Arm. "I never had any exposure really prior to this in any major construction sites or anything like that," he says. "The expectations I had coming, they had to change. I guess I had visions of the old Dodge City days—knock them down and drag them out. That didn't materialize." He was "pleasantly surprised" by construction camp life and gladly discarded his preconceived ideas, settling into largely preventative security duties.

Metropol was awarded the security contract for the Bull Arm site in 1991. Over the course of Hibernia's construction period, uniformed security guards checked thousands of identification cards at the main gate entrance and patrolled the site during 12-hour shifts.

For some of the workers, Hibernia is a frontier project in more ways than one. Lisa Whelan of Placentia had worked at a variety of jobs, from waitress to lawn bowling instructor. But she has always enjoyed working with her hands, and was taking a carpentry course when she learned that HMDC was offering a course in working with rebar for women. Although it was difficult to get admission into the course, Lisa saw an opportunity that she could not allow to pass her by and persisted until she joined the group of women trained at Newfoundland Steel in Argentia in the non-traditional trade. She started rebar work at Bull Arm in the summer of 1993, placing steel for the GBS's outer ice wall and supporting XVT wall. "You get used to the work," she says, "and it feels good when you put in a good day."

Whelan eventually became the first female foreperson at the site. In the spring of 1996 she was eagerly anticipating the mating of the GBS with the Topsides, declaring, "That's going to be a real joy for me to watch that because I know that I was a part of it from the bottom right up to the top."

1992

D52
Hibernia

For Bob Warren of Chance Cove, the decision to make the transition from fisherman to Fisheries Liaison Officer for the Hibernia Project was not an easy one. "My wife stayed awake into the nighttime saying, 'What the hell are you doing, Bob, going down there? You're doing okay fishing.' I had a problem making up my mind to come here because I was used to fishing, I made a decent living fishing, and now I was going at something where I wasn't exactly sure what I was getting into," he recalls, "But it worked out great." It's been "a different ballgame" for Warren, who has enjoyed being part of the team that put Hibernia together and remembers the day that the GBS floated in the dry dock with particular satisfaction. "There were a lot of doubts," he says, "and 'Newfoundlanders can't do this or can't do something else.' Now we've proven to the world that we can do it."

Cartoonist Kevin Tobin's salute to the Newfoundland labour and determination that helped Hibernia to achieve a major milestone, the floating of the GBS to the deepwater site in November of 1994.

Many workers share Warren's sentiments, believing that Hibernia has given Newfoundlanders, who held the majority of the positions at Bull Arm, a very public forum to demonstrate their adaptability and competency in a new industry. They see this as an opportunity to encourage the development of similar projects for the benefit of the regional economy, and also to offer an alternative view to the perception of the province and its people as chronically dependent on government assistance, a view that seems to have persisted in some form or other for decades.

The Atlantic Accord signed by Premier Brian Peckford and Prime Minister Brian Mulroney in February of 1985 provided for joint federal-provincial management of offshore petroleum resources and recognized the right of Newfoundland and Labrador to be the chief beneficiary of these resources. Coming after the March 1984 ruling by the Supreme Court of Canada that rights to explore and exploit resources in the Continental Shelf in the area of the Hibernia Field fall under jurisdiction of the federal government, not under Newfoundland's provincial jurisdiction, the Accord was lauded as finally recognizing Newfoundland as an equal partner in Confederation. But for Newfoundlanders working on Hibernia the successful completion of the project in the face of detraction was more satisfying than any number of declarations recorded in legalese.

Courtesy of *The Evening Telegram*

The Atlantic Accord was signed on February 11, 1985 at the Hotel Newfoundland in St. John's. Pictured left to right: Pat Carney (then Energy Minister), John Crosbie (then federal Justice Minister), Brian Mulroney (then Prime Minister), Brian Peckford (then Premier of Newfoundland and Labrador, and Bill Marshall (then provincial minister responsible for the offshore).

Some workers point to the fickleness of media coverage of the project, throwing bouquets when deemed appropriate, and otherwise latching onto information about problems or setbacks with apparent relish. Gulf Canada Resources announced that it was withdrawing its quarter-share partnership in the project on February 4, 1992 because of difficulties experienced by its controlling shareholders. John Crosbie responded to a *Globe and Mail* editorial reacting to the withdrawal entitled "A Happy Ending to Hibernia, and soon" with a letter printed in the paper on January 4, 1993 charging, "Your central Canadian condescension and insularity has never been more evident." Crosbie criticized the coldness of the editorial and asserted that Hibernia offers a viable alternative for a population that had been devastated by the loss of thousands of fishery jobs, and is a move in the direction of "a self-supporting Newfoundland." Furthermore, he claimed, "It would end the endless smug assumptions and put-downs that emanate unceasingly from people such as you who assume that 'national' means Toronto and that all real progress can only occur within the area encompassed within your own eyesight."

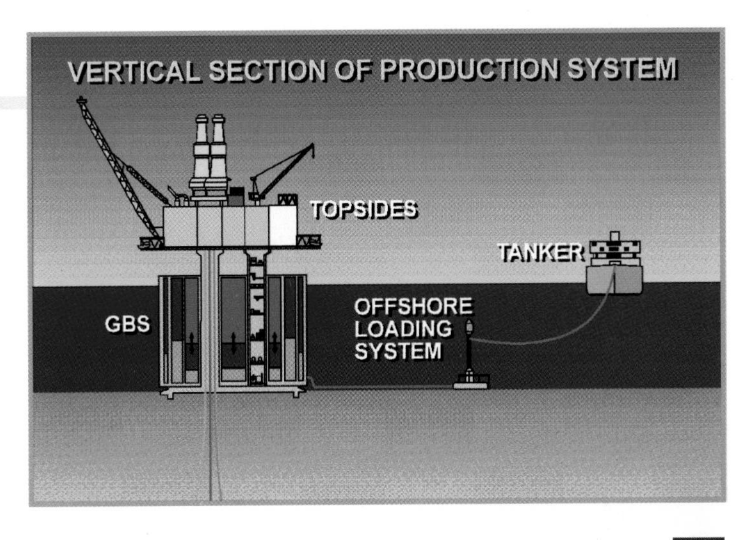

The main components of the Hibernia Production System are the Gravity Base Structure (GBS), Topsides, Offshore Loading System (OLS), and the crude oil transport tankers. Oil from the GBS storage cells reaches the double-hulled tankers via the OLS loading hose, which is connected to a vertical pipe protruding from a base mounted on the ocean floor.

"I remember vividly in June of 1991 looking at the job we had to do on the drawing and knowing the effort to build one of these things that I've gone through in the UK, and how long it has taken those in the UK to start producing the right quality of work. And I'm so astounded that we've been able to get the quality job that we've had."

— Tony Hylton, Topsides Site Manager

PRESTRESS
IRON WORKERS
LOCAL 764
PRESTRESS
DANNY

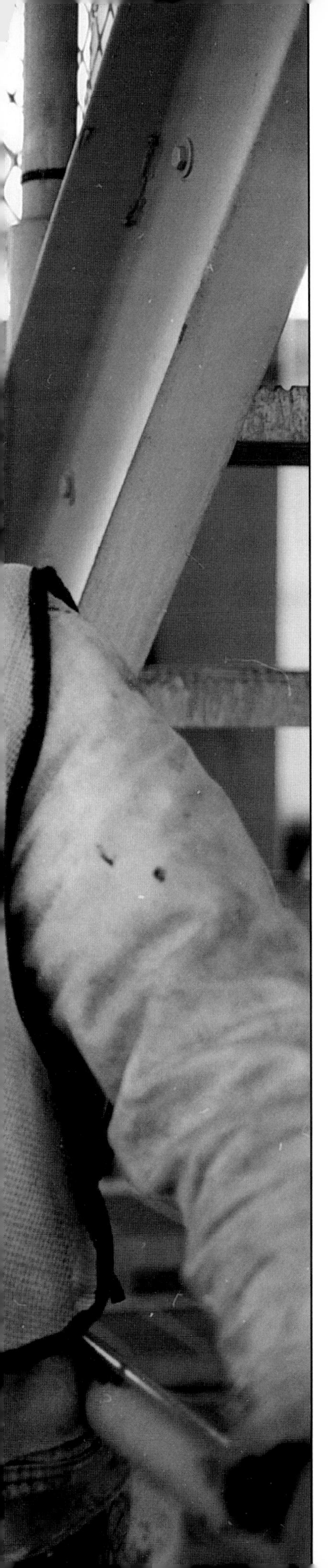

"The experience that I have received here, you probably wouldn't get anywhere else. At Bull Arm you are exposed to an offshore atmosphere without being offshore."

— Craig Ryan, Mosquito Cove Fire Department Chief, January 1996

Gulf's withdrawal resulted in an expenditure slow-down, and the project was delayed by about a year. After a worldwide search and much tension, the news was released that the lost Gulf investment would be replaced by increased support from partners Mobil Oil Canada and Chevron Canada Resources, joined by the Government of Canada and Murphy Oil Corporation. The consortium's other member, Petro-Canada, retained its original interest level, although it sold a small piece of its stake in 1996, during which year Norsk Hydro joined with a five percent interest.

But the period of uncertainty that preceded the announcement and the way it was handled by the media and reacted to by the general public has left a lasting impression with many of those who have personal investments in Hibernia. "The project has had some bad moments. It's had a lot of negative press, the Gulf pull-out. A year and a half after the Gulf pull-out it was still pretty damned miserable," recalls Topsides Site Manager Tony Hylton, adding, "We've overcome those and all the technical problems that we had."

Juan Beckett, a rodman from Gambo who started work at Bull Arm in the spring of 1993, also remembers this period and shares the satisfaction of many of his coworkers in its resolution. "You always hear more about this place when it was doing poorly than when it started to go well. When we first started here there were people constantly saying that it wasn't viable and wasn't going to be built," he said in June of 1996, "But it is getting built and it's good quality."

While the project has certainly had its detractors, there has also been ample positive feedback from the public. Guided bus tours of the Bull Arm Site have tried to accommodate as many visitors as possible, offering impressive views of the GBS and Topsides and answering questions. Between the introduction of the tours in late 1992 and the final one in October of 1996, more than 65,000 people, ranging from curious tourists to senior government and business representatives from countries worldwide, have visited the site.

Pipefitter John Power of Grand Falls-Windsor wasn't even in Newfoundland when he welcomed inquiries about Hibernia. He has worked in Nova Scotia and the Northwest Territories and was about to return north for employment when, three days before his planned departure in August of 1994, a call came from his union informing him that he had been given a position with the PASSB joint venture working on the Hibernia Topsides. During some time off in 1995, Power travelled to Prince Edward Island to watch his son play hockey. "The kids think it's great for Dad to be working on Hibernia," he says, and his son mentioned his father's job. "I was asked so many questions by the local residents of PEI there in the arena when they found out," Power recalls, "And there were so many questions coming at me, and I felt so proud to say that I was part of this huge project, that so many people outside of the province and around Canada would ask all those questions. They were amazed that the huge project was going ahead. And, you know, I feel darn good about being able to answer a lot of those questions, and told them some of what it's like to be a part of it, and what kinds of amazing things they would see in a project such as this. So many things would come to mind."

"I've had opportunities to work outside of Newfoundland but have chosen to stay in the province, and I must say that it's worked out great."

— David Day, Environment, Safety and Quality Advisor

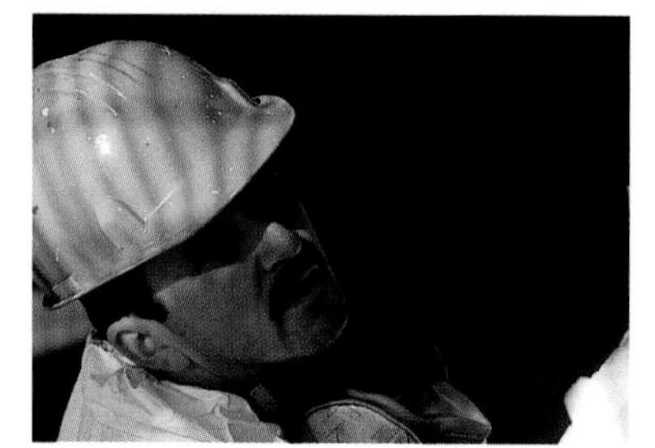

Few Newfoundland workers who are asked about their experiences at Bull Arm neglect to mention the contributions that "Come from Aways" (CFAs) have made to the project. It has indeed been a multinational effort, as Robert Goodyear, a rodman and former fisherman from Ladle Cove, Notre Dame Bay, observed. "Basically," he says, "there are people from all over the world here. Our superintendent is from Croatia, and I have a foreman from Bosnia." He also points out the participation of people from Norway, France, and the United States—"people from all corners of the world."

GENEROSITY

The Hibernia Union Tradespeople, Bull Arm Site, representing sixteen unions, donated $1.5 million to the General Hospital Foundation over the period from August 1994 to June 1996. The money was for the purchase of a Catheterization Laboratory, a diagnostic and treatment tool for cardiac patients to be located at the General Hospital, Health Sciences Centre in St. John's.

"A lot of possibility will always be here at Bull Arm, but you don't build things with just a facility. It's the people that are running things and doing the work."

— Harald Gulaker, Construction Manager of the Topsides Hook-Up Team.

Norwegian Harald Gulaker, Construction Manager of the Topsides Hook-Up Team, worked for about 13 years in the North Sea offshore industry and thinks that the growing pains of Hibernia's early stages are very much like what he has witnessed there. "The early days that we went through were, more or less, what Newfoundlanders are going through now with the offshore industry here," he says. Gulaker adds that Norway and Newfoundland have a similar diet and climate, as well as small communities with a connection to the sea: "We have been dependent historically on a lot of fishing, as you have been here."

Astor Nyborg of Stavanger, Norway, is the GBS Construction Manager. He joined the project in 1994 as part of the joint venture responsible for overseeing GBS construction, and has found Newfoundlanders to be "strong workers": "I have a deep respect for them." Nyborg has also found cultural similarities between his native country and Newfoundland, so that the adjustment to living in the province has not been difficult. Rather, the challenge of his managerial position has been overseeing a complex construction project in a multi-national worksite. "I have people here from all cultures, different styles. Of course, my biggest challenge is to try to find the best in everyone," he says, "I learn a lot because I meet a lot of people who look into problems with completely different eyes."

G.M.C.

IRON WORKERS
LOCAL 764
WAYNE
WAYNE

Administrative Assistant Janet Lahey, originally from Bell Island, points to the technology transfer and exchange of ideas and methods that such a situation facilitates, and has been very impressed by the cooperative effort. "After you've been here a while you forget that this person is from Holland, that person is from Norway, and this person is from Arnold's Cove," says Lahey, who began work at Bull Arm in 1992. "You all have your work and work together to get it done to the best of everyone's ability. That's what's nice: everybody is coming together for the good of one thing."

Sandy Cornelius, Well Performance Team Lead with Hibernia, echoes Lahey's sentiments. "We seem to have a team-based organization that is structured deliberately to force interdependencies, so that you need to link arms with others to succeed," he says. Cornelius is from Kilwinning, Scotland and has worked at projects ranging from designing synthetic tobacco to being a design engineer in the nuclear power industry. Drawn to the challenges of a major start-up project, he came to Newfoundland in 1996 to work on a secondment with Hibernia and became a member of several of the development's work teams.

Civil Works Manufacturing Data Books Coordinator/Project Certification Coordinator Colette Leonard from St. John's also emphasizes the cooperative effort of the massive project. "It's surely been a team effort," she says, "You can't build a project like this without working as a team. It requires coordination and cooperation between every discipline and every group, whether it's a guy telling a crane to move over and bring the rebar out to the barge, or the engineer in town who's designing what size it is. It doesn't matter what level. Every little bit has to come together at the right time. That is, I think, the biggest key to making something like this a big success."

Among other things that keep her schedule full, Leonard's positions dictated that her contribution to the effort include independently reviewing inspectors' reports, and she found that every day brought a new work issue to her attention. Leonard began her secondary education at Memorial University before transferring to and graduating from the University of South Florida, in which state she then worked for several years before moving on to a consulting firm in Toronto for a year. She was happy to be back in Newfoundland as part of the Hibernia team. "I'm just so thankful to be here, to have the opportunity to have learned so much," she says, "I could have worked 15, 30 years and not have seen and done what I have here. And what I've learned here is going to be really beneficial to my working on any similar type project, whether it's building a GBS or working with the production company after it's been built."

COLLETTE

SURE LOCK
SURE LOCK

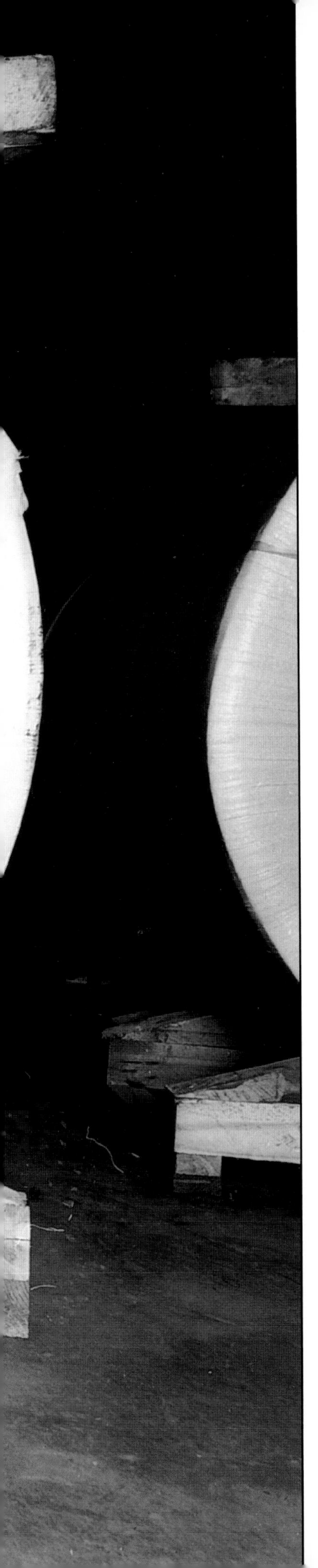

Tony Rumboldt of Rocky Harbour, a carpenter at the Hibernia construction site, placed second in the Canadian National Carpentry Apprenticeship Contest, sponsored by the Carpenter's Union, in 1995.

Some Hibernia employees fear that the contributions of Newfoundlanders to the project have been overshadowed by debate about the fabrication of supermodules in Korea and Italy or by the visibility of "CFAs" in managerial positions. As Assistant GBS Construction Manager, Calgary native David Sverre occupied such a senior level position, and has been pleased with the Canadian participation in the enterprise. But he expresses concern that "somehow that average person who has been out here and put the effort in hasn't been given the credit that they're due." In actual fact, more than 90 percent of the 5800 people working at Bull Arm during peak construction activity were Newfoundlanders.

Sverre zeroes in on the participation of Newfoundlanders in particular by giving the example of the cost control group. He recalls a team meeting with the group: "I went around the whole group and I said, 'Okay, tell us who you are, what you do, and I also want you to tell me where you're from specifically.' And virtually every one of the working people in the group was from the Avalon Peninsula of Newfoundland, and here they are. And I said, 'Do you realize that this group, controlling a 6.2 billion dollar job—the biggest construction job in the world —are virtually 100 percent from the Avalon, as are so many of the workers, the people who are actually doing it, the hands-on cost engineers and everything."

The M-20 Wellhead Module, one of five supermodules that comprise the Topsides portion of the Hibernia platform, is another major part of the Hibernia success story about Newfoundland workers. 1200 Newfoundlanders helped fabricate the 5000 tonne module between December 1992 and March 1995. It contains the platform's blowout prevention system, and piping and wellheads for the wells that will produce the oil. It also contains what are known as Xmas trees: assemblies of control valves and pressure gauges shaped like evergreen trees, installed on top of the wellhead to control oil and gas flow during production. A ceremony was held in April of 1995 in the Topsides Module Hall to mark the structural completeness of this supermodule, and it was rolled out of the hall on 808 wheels and onto the Topsides Assembly Pier at the end of that month, with all the pomp and cheering befitting the occasion.

Comprised of five supermodules and seven mounted structures, the Topsides portion of the Hibernia platform united skilled workmanship from Italy, South Korea, and Canada. These facilities, shown here on the assembly pier, will rest on the GBS and contain drilling, production, and processing equipment, as well as personnel accommodations.

NEPTUN

NEPTUN

Preparing to assume the position of HMDC President on October 1, 1996, New Brunswick native Harvey Smith prophesied, "It's going to be exciting times as we work over the next couple of years." Indeed, just one month to the day that he assumed his new position, Smith witnessed the culmination of six years of design, engineering and construction with the completion of the GBS and Topsides, which he declared "undoubtedly our biggest achievement to date."

The Mating: Joining the colossal GBS and Topsides structures into an integrated unit demanded power and precision. Two massive barges guided by tugboats ferried the Topsides to the GBS mating site for a distance of 1.5 kilometres. It was then maneuvered over the ballasted GBS. The GBS was deballasted and rose to meet the Topsides, and the pair were lined up to within one millimetre of the target contact point using a jacking system.

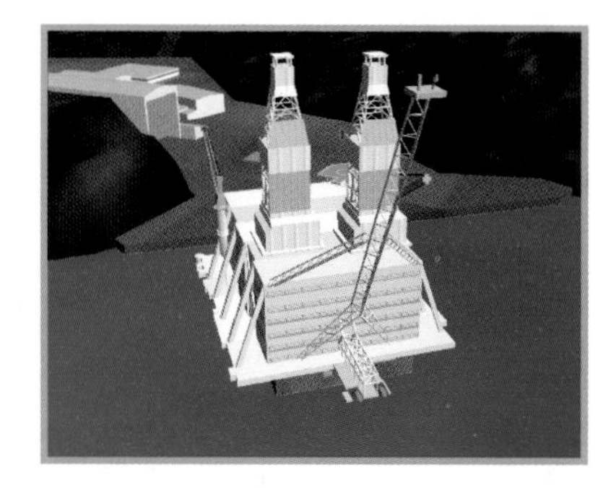

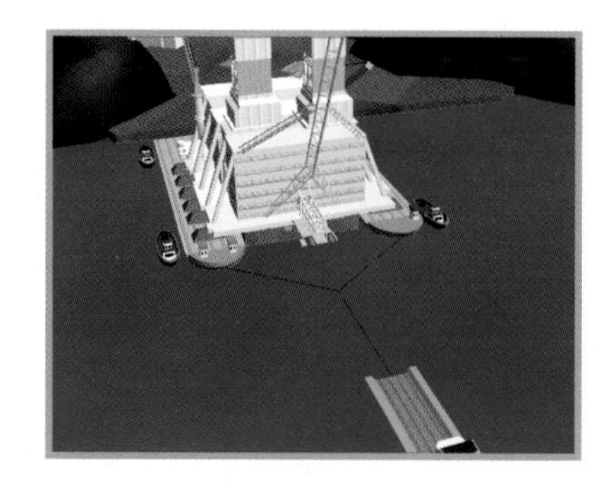

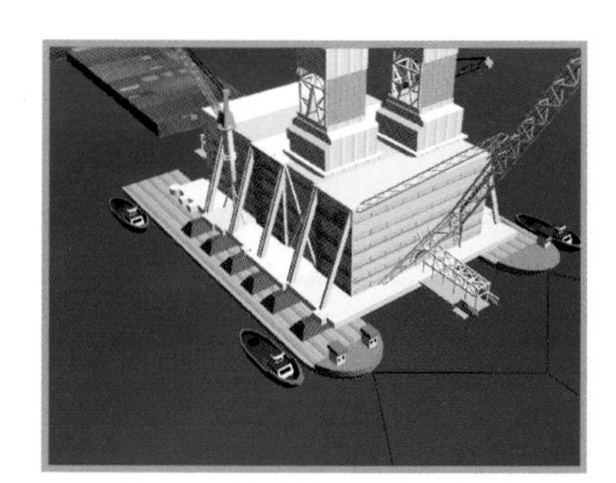

NEPTUN

There was an even more exciting milestone to celebrate before long. Unfavourable weather conditions held up Topsides to GBS mating operations for several days, but the two colossal structures were joined as an integrated unit at Bull Arm in the early hours of February 28, 1997. The feat began on the preceding evening, when journalists and Bull Arm workers gathered near the pier to watch the start of the towing of the Topsides to the deepwater site. Roland Currie, a welder with Hibernia since 1993, had his camera on hand "to get a few pictures for memento's sake." He recalled that when he started work on the Topsides "it was just the pieces of module and the grillage underneath," and was delighted to witness the progress. "This is a good thing to see after all the work that was put into it," he said, "Now you can actually see what you started out to do. It's good to finally see it off."

Even Construction General Manager Henk van Zante was impressed with the relative quickness of the mating procedures, and Oyvind Hagen, Marine Operations Manager with Norwegian Contractors, proudly reported that everything had gone as planned, with the massive structures mated to within one millimetre of the target contact point. The completed platform is an engineering marvel standing 224 metres high.

Everyone involved in the 6.2 billion dollar Hibernia project, whether from the Avalon Peninsula of Newfoundland, Labrador, New Brunswick, Norway, or France, looked excitedly forward to its penultimate milestone, the towing of the mated GBS and Topsides to the Hibernia Field. In 1996, Camp Maintenance Foreman Don Best, originally from Come by Chance and a long-time resident of Clarenville, said, "When June of '97 rolls around, that's going to be one big day!" Site Security Supervisor Jim MacDonald agreed. "We'll be right there to the bitter end," he declared, "We'll have a firsthand view of it all. It will be quite a day for sure."

The christening of the Hibernia Platform on May 9, 1997 by Aline Chretien, wife of Prime Minister Jean Chretien, was delayed by protesting fishery workers who gathered at the entrance to the Bull Arm site.

Best and MacDonald could not have foreseen that the tow-out would actually begin ahead of schedule in May of 1997. But in the early morning hours of Friday, May 23 a diving crew working 74 metres beneath the ocean surface at Bull Arm and employing underwater torch technology finished the task of cutting the six huge mooring chains holding the platform in place. Ready for its 500-kilometre trip to the Grand Banks installation location, the Hibernia Platform was pulled away from its deepwater site and into the fog by an entourage of powerful tugboats. Newfoundlanders flocked to the coast to catch a glimpse or take a photograph of the towering one-of-a-kind structure on its way to the edge of the Grand Banks to its pioneering position in the Hibernia Oilfield.

Henk van Zante, General Manager of Construction, is a seasoned veteran who was knighted by Queen Beatrix of the Netherlands for his contributions to the offshore oil industry worldwide.

Ten powerful tugboats assembled around the Hibernia Platform in a star configuration in preparation for tow-out to its offshore production site. The platform was deballasted, leaving 73 metres of the GBS underwater and allowing a sufficient amount of clearance between the bottom of the platform and the sea floor. The six huge mooring chains were cut and the tugboats realigned to tow the platform into Trinity Bay and eventually into position above the Grand Banks installation site. The structure was ballasted until it descended to and penetrated the ocean floor, and was stabilized there by a grout mixture of cement and sea water. Shortly after, workers began pumping more than 400,000 tonnes of iron ore ballast into the GBS.

A support vessel tows a "lassoed" iceberg well out of the path of the Hibernia platform and its entourage of tugs enroute to the Grand Banks production site.

Hibernia workers hold proud memories of the day that Archimedes Principle pulled through and the GBS floated for the first time in October of 1994, and of watching the 5000-tonne M20 Wellhead Module roll out of the module hall on April 30, 1995 after about five million person-hours of work. These events had conditioned them for the tow-out, but that day was one of mixed emotions, and for many of the workers marked the end of their involvement with the Hibernia Project. What remains besides memories is new expertise, a wealth of experience, and the awesomely tangible evidence of their labour, skill, and effort: a 1.2 million oil production platform sitting in the Atlantic Ocean, poised to penetrate the Hibernia and Avalon Sandstones for first oil, marking their tremendous pioneering success.

"I" for the icebergs come drifting down through,
"J" is for the job that we're out here to do,
"K" is for the knowledge that built this damn thing,
And "L" is for lifting what the supply boats bring.

— From "The Rig Workers' Alphabet." recorded on *Wave Over Wave: Old and New Songs of Atlantic Canada*, Jim Payne and Fergus O'Byrne, SingSong, SS 50440. Copyright Jim Payne, SOCAN

Hibernia Milestones

Telegra

Monday, July 18, 1988

Single copy 40 cents — Home delivery $2.30 a w

HIBERNIA DEAL SIGNED T

Deal so good it's "star

Newfoundl

in biggest decision says Pec

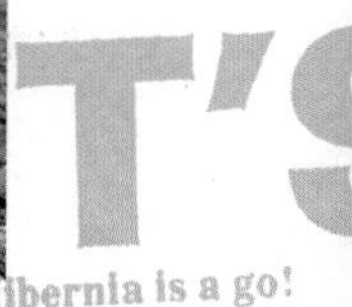

1965 First seismic surveys conducted.

1966 First exploration well drilled offshore Newfoundland.

September 1979
Hibernia discovery well drilled.

September 1990
Announcement of Hibernia Agreement. Gravity Base Structure (GBS) contract awarded. Topsides engineering and procurement contract awarded.

October 1990 Work began at Bull Arm.

September 1991
M-20 Wellhead module contract awarded.

September 1992
Commenced GBS construction in the dry dock.

A GO FOR HIBERNIA!

nt, the Government of Canada Mobil Oil Canada Ltd. have developing the $5.6-billion oil- Grand Banks, 315 kilometres

Jake Epp, Canada's Minister of Energy, Mines and Resources, said the

Rex Gibbons and chief executive officers of the oil companies.

"This day has been a long time coming but few will have reason to doubt that it has been well worth the wait. Today a major milestone has been reached," said Mr. Epp during a news conference announcing the major project agreement.

After 11 years of political wrangling, court battles, political manoeuvring, hopes being built up and then dashed, considerable money lost by early investors and the heartaches, this morning's announcement was sweet music indeed to people who filled the hotel ballroom from virtually every sector of society in the province.

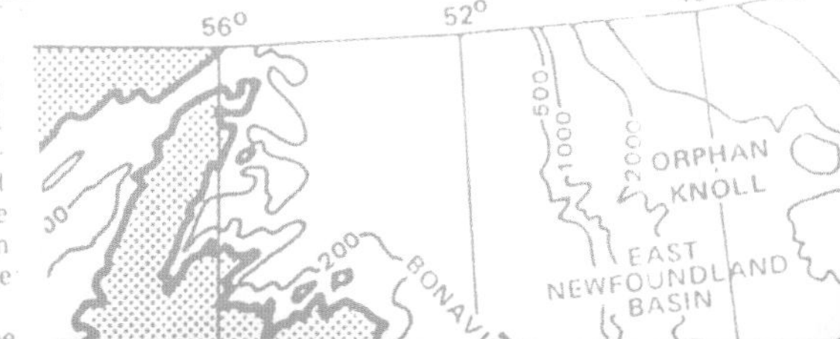

November 1992 Commenced fabrication of the M-20 Wellhead module.

January 1993 Contracts awarded for the M-10 Process Module and M-50 Living Quarters/ Services Module.

February 1993 Contracts awarded for the M-30 Mud Module and M-40 Utility Module.

April 1993 First ceremonial concrete pour for the GBS.

August 1994 GBS slipforming commenced in the dry dock.

October 1994 Dry dock flooded and the GBS floated.

November 1994 GBS towed to the deep water construction site (DWS).

Hibernia project on a ro

Hibernia

milestone

By GLE...
The Even...

...ost Newfou...
...nd by and...
...hed or a boa...
...y without pit...
...ut workers
...Bull Arm or...
...nd and wate...
...torey high
...s rolled onto
...just two me...
...s of a hy...
...n 808 wheels...
...It's great to...
...module ha...
...king on it,
...d, a welding
...site.
...ne worker
...e by it took...
...logs to mov...
...ance.

...he M-20 m...
...foundlande...
...first of five supermodules that
...make up the topsides portion
...he massive Hibernia platform.
...M-20 is 86 metres long, 22
...res high and 18 metres wide.
...This is a very important mile-
...for us, obviously," said Ken

building the structure.
"I think a lot of people here will
take away now the...
learned...

...esents
...million
...k that
...ried o...
...coupl...
...so it's
...y for us
...I think
...proud
...for the
...rkers."

...resident
...en Hull

...out five mil-
lion hours of work that's been carried on over the last couple of
years here, so it's a proud day for
us and I think particularly a proud
day for the workers."

Hull said the Newfoundland
workforce has acquired experi-
...technology and new weld-

workforce th...

50¢ + 4¢ GST

Hibernia ...es milestone...leb complete

...GBS caisson wi...
...d to a height of...
...en the construct...
...hafts will contin...
...em above the c...
...ing the complete...
...s maximum he...

...e GBS will then b...
...topsides oil prod...
...ore the platform...
...produc...

Home sweet home: GBS set

By CHRIS FLANAGAN
Business Editor

The Hibernia platform began its slow crawl over the last 10 ...

January 1995 First shipment of Utility Shaft decks arrived from Nova Scotia.

April 1995 Second shipment of Utility Shaft decks arrived. M-20 Wellhead Module moved to the Assembly Pier.

May 1995 M-30 Mud Module arrived from Italy. M-10 Process Module arrived from South Korea.

June 1995 M-40 Utility Module arrived from Italy. M-50 Living Quarters/Services Module arrived from South Korea.

July 1995 Flare Boom, two Lifeboat Stations and cranes installed on the Topsides.

August 1995 Final shipment of GBS Utility Shaft decks arrived. Helideck installed on the topsides.

Hibernia Milestones

Mating praise flows at Bull Arm

Tobin calls production platform an engineering marvel

By BERNIE B...
The Evening T...

...took an ancient
...nb, to get the per...
...Hibernia topside w...
...ase structure (G...
...ster which sits in...
...specimen of...
...us.

...e first major step...
...roduction was a...
...the mating and th...
...orated Monday w...
...y tossing accolad...
...esponsible for th...
...evement, mainly
...foundlanders.

...'s one of the truly great engi-
...ng marvels of the world," said
...ier Brian Tobin to the senior
...nia staff and about 200 invited
...from the engineering, con-
...on, academic and business
...e nity of St. John's at the Bull
...ll P.

...successful mating is an
...le feat and a credit to the
...ous workers of
...the lland and Labrador."
...tion : completion of the pro-
...rest said the men and women
...d on the project sent a

ing in

November 1995 Piperack Structure installed on the Topsides.

December 1995 Two Drilling Derricks installed on the Topsides.

June 1996 Final placement of Mechanical Outfitting in the GBS Utility Shaft.

November 1996 GBS and Topsides complete.

March 1997 Mating of the GBS and Topsides.

June 1997 Towout to offshore oil production site.

August 1997 Drilling scheduled to commence.

December 1997 Oil production scheduled to begin.

Photo by Marc Pike

Ned Pratt was born in St. John's, Newfoundland in 1964. He grew up in the small community of St. Catherines at the head of St. Mary's Bay. The predominantly Irish area of Newfoundland is known for its gentle beauty and challenging way of life. In 1986, Pratt graduated with a BA in Art History from Acadia University in Nova Scotia. Pursuing his interest in the arts, Pratt went on to the Nova Scotia College of Art and Design where he earned a Bachelors Degree in Fine Arts.

Upon graduation, Pratt began a freelance career in photography, shooting assignments as photo editor for the *Sunday Express*, a critically acclaimed newspaper published in Newfoundland between 1989 and 1991. In subsequent years Pratt's work appeared in the *New York Times*, *Newsweek*, *Maclean's*, *The Globe & Mail*, *Canadian Geographic* and *The Financial Post*.

Throughout the course of his freelance career, Pratt has worked on diverse projects in challenging environments. A wide variety of clients and broad spectrum of experience have helped him hone both technical and creative skills. His portfolio includes commercial, fashion, portrait, and food photography for numerous national magazines and corporations.

Pratt is perhaps best known for his mastery of black and white photography and the art of portraiture. His work has been showcased in a number of national exhibitions and most recently with G. Ray Hawkins Gallery, Santa Monica, California in a collection entitled *Wedding Days: Images of Matrimony in Photography*. The show is currently touring the united States and Japan.

Pratt's photographs can be found in various private, public and corporate collection including the Ford of Canada Photographic Collection and the Canadian Museum of Contemporary Photography. Publications include *Rabbit Ravioli* (1994), *Faces of Canada* (1992) and *Unholy Orders* (1992).

Photo by Keith Gosse

Greg Locke started his photojournalism career in the early '80s in St. John's after studying anthropology at Memorial University, getting a pilot's license, and knocking around the offshore oil exploration on the Grand Banks.

Since then he has traveled to more than fifty countries on four continents for such publications such as *Maclean's*, *Equinox*, *Time*, *Bunte*, *der Spiegel*, *BusinessWeek*, *Canadian Geographic*, Reuters News Pictures and various national and international magazines and newspapers.

Today he is based in St. John's where he lives with his wife and daughter and is a regular contributor to *Maclean's* and Reuters as well as documenting the Hibernia oil project for various publications and oil companies.

The Hibernia project has kept Greg busy in St. John's but he still managed to complete assignments in Bosnia, Serbia, Croatia, Haiti, Rwanda, Zaire, Kenya and Somalia in 1996 and will be publishing a second book this year about his work in Africa with the international medical relief agency, Medecins Sans Frontieres.

Although this book project started only three years ago, Greg began documenting the Hibernia construction project when the first surveyors and construction crews entered Mosquito Cove and Bull Arm in 1990.

NOBLE
EDDIE PAUL
NOBLE
Noble Drilling (Canada) Ltd. is committed to provide personnel, processes and equipment to support drilling and drilling related activities on Canada's East Coast. Our specialty is offshore drilling and maintenance of drilling systems. Our goal is to become the leading offshore drilling contractor on Canada's East Coast.

CAHILL State

A JOINT VENTURE

G. J. Cahill and Company and the State Group formed their joint venture in early 1991 to pursue work on the Hibernia Project. The partnership was successful and its members were proud to have been able to participate significantly in the construction of the Hibernia platform. Over the six years of construction the joint venture carried out over 500,000 person hours of work on the project. Some of the contracts that the joint venture carried out on the project included:

- Electrical work for various Bull Arm support facilities including the main contract for all the Topside fabrication facilities
- Electrical and instrumentation work for the utility shaft modules of the Gravity Base Structure assembled in Dartmouth, Nova Scotia
- Electrical and instrumentation work for the installation of all electrical and control systems on the Gravity Base Structure at Bull Arm, Newfoundland

G. J. Cahill and Company is an electrical and instrumentation construction company based in St. John's, Newfoundland with offices in Dartmouth, Nova Scotia. Cahill has carried out significant projects in each of the four Atlantic provinces. G. J. Cahill was the project sponsor of the joint venture.

The State Group Limited

The State Group is a national multi-discipline construction company based in Toronto, Ontario with offices in Quebec (National State), Ontario, Manitoba and British Columbia. The State Group is a division of the Bracknell Corporation, a publically traded Canadian company.

Photo courtesy of HMDC

Photo courtesy of HMDC

GOVERNMENT OF
NEWFOUNDLAND AND LABRADOR

Message from the Minister of Mines and Energy

The completion of the construction of the Gravity Based Platform for the Hibernia oil field represents an important milestone in our oil and gas industry and in the history of the province.

This massive construction project demanded a highly skilled labour force, competitive industrial infrastructure, and an active high technology community.

At peak construction, approximately 5700 people worked at the Bull Arm construction site. The quality of work completed rates second to none anywhere in the world. By the end of 1996, over 61 million person-hours of work were completed, of which 65 per cent or almost 40 million person-hours, were completed by residents of this province.

This platform represents more than a construction project and an engineering wonder—it is a source of pride. Our workforce met the challenges of constructing the huge platform and positioned Newfoundland and Labrador at the leading edge of offshore petroleum technology.

Participants in the project received significant technology transfer and acquired skills that will enable them to compete worldwide. The Hibernia Construction Project also laid the foundation for other offshore developments that will further enhance the province's economy and ability to take its place in the global oil industry.

As the Minister responsible for Mines and Energy in the province, I am proud to have been a part of the Hibernia Project since it began in 1990. Also, on behalf of the Government of Newfoundland and Labrador, I extend my congratulations to the men and women whose efforts made the construction of the Gravity Based Platform a success.

Rex Gibbons, Ph.D, P.Geo

Photo courtesy of HMDC

Photo courtesy of HMDC